KU-626-436

The Institute of Biology's
Studies in Biology no. 32

Fungal Saprophytism

Second Edition

Harry J. Hudson
Ph.D.
University Lecturer in Botany and Fellow
of Fitzwilliam College, Cambridge

Edward Arnold

First published 1972
by Edward Arnold (Publishers) Limited,
41 Bedford Square,
London, WC1B 3DQ

Reprinted 1976
Reprinted 1977
Second Edition 1980

British Library Cataloguing in Publication Data

Hudson, Harry James
Fungal saprophytism. – 2nd ed. – (Institute of Biology. Studies in biology; no. 32 ISSN 0537-9024).
1. Fungi – Ecology
2. Saprophytism
I. Title II. Series
589'.2'045 QK603

ISBN 0-7131-2792-9

Printed and bound in Great Britain at
The Camelot Press Ltd, Southampton

General Preface to the Series

Because it is no longer possible for one textbook to cover the whole field of biology while remaining sufficiently up to date the Institute of Biology has sponsored this series so that teachers and students can learn about significant developments. The enthusiastic acceptance of 'Studies in Biology' shows that the books are providing authoritative views of biological topics.

The features of the series include the attention given to methods, the selected list of books for further reading and, wherever possible, suggestions for practical work.

Readers' comments will be welcomed by the Education Officer of the Institute.

1980

Institute of Biology
41 Queen's Gate
London SW7 5HU

Preface to the Second Edition

Saprophytic fungi are indispensable for the maintenance of the carbon and mineral cycles in nature and are particularly important in the utilization of cellulosic materials which form the bulk of decomposing plant remains. As a group they can grow in almost all conceivable habitats in which some form of organic carbon is available. They occur in aquatic environments both fresh water and marine. Some even prefer to grow in much more concentrated solutions such as 40–60% sucrose or in very dry environments. They grow from −6 to 60°C, on refrigerated foods and on microbially self-heating composts.

Although much is known about those of economic importance such as wood-degrading fungi, there is a dearth of information about many of the others. It is only in the last few decades that studies have been made on fungal ecology. There is adequate scope here for fieldwork and laboratory experimentation. This booklet deals with fungal saprophytes with a bias towards the ecological approach. It omits to cover studies on soil fungi but many of the broad concepts introduced apply equally to these. No attempt has been made to give any taxonomic treatment of the fungi. The scheme followed is that used by WEBSTER (1970).

Cambridge, 1979

H. J. H.

Contents

1 Nutrition of Saprophytic Fungi

1.1 Saprophytism and parasitism

Fungi are heterotrophic for carbon compounds. They are unable to use carbon dioxide as their sole carbon source as green plants can. They require complex organic compounds synthesized by other organisms, especially autotrophic seed plants. Saprophytic fungi either colonize dead plant and animal remains or they absorb organic materials which have exuded or leaked from living or dead organisms. The distinction between saprophytism and parasitism by fungi cannot always be clearly demarcated. Many parasitic fungi can also live as saprophytes either on the host which they have killed or on other dead organisms. The ability to invade living tissues distinguishes them from true saprophytes. They are considered as facultative parasites as distinct from obligate parasites which can only grow and fully develop in association with an appropriate living host. Most saprophytes and facultative parasites can be grown readily in culture whereas it is only recently that a number of the obligate parasites have been cultured axenically.

Many fungi may switch from one type of nutrition to another according to circumstances. A good example is the Basidiomycete, *Armillaria mellea*, the honey fungus, a devastating root parasite of forest trees. After killing its host, it lives on as a saprophyte utilizing the carbon and nitrogen sources in the dead trunk and roots. In addition it can also enter into a balanced mycorrhizal association with some orchids, such as *Gastrodia elata*. Similarly in many green orchids the *Rhizoctonia* endophytes are capable of maintaining themselves indefinitely in the soil either as vigorous saprophytes on cellulosic materials or as facultative parasites causing 'damping off' diseases of seedlings.

This text is concerned with saprophytic fungi which can go through their whole life cycle utilizing dead organic materials, although reference will be made to facultative parasitism where parasitism is an advantage in prior colonization of a particular food base or substratum.

1.2 Carbon sources

Simple sugars are the most readily utilized carbon sources and may serve as the sole source for the majority of fungi. Glucose is utilized by virtually all fungi and, for most, fructose and mannose are equivalent. The only exceptions to this are found in one odd order of Mastigomycetes, the Leptomitales, which includes fungi, such as *Leptomitus lacteus*, that are incapable of utilizing hexoses and other sugars; they grow on acetates and fatty acids in water polluted by sewage.

Xylose is the most generally utilizable of the pentoses and may be superior to glucose for some fungi. Sorbose is anomalous in that it supports good growth of very few fungi and if present may inhibit the utilization of other sugars which alone are readily utilized. It is definitely toxic to others, a property which enables it to be used as a suppressant to retard lateral spread of hyphae in culture.

Sucrose, the characteristic disaccharide of seed plants, is a good source of carbon but is not so universally available as is maltose. Before these can be utilized the fungus must produce the necessary extracellular hydrolytic enzymes to split the disaccharides into their component monosaccharides for absorption. Some Chytridiales, Mucorales, such as *Rhizopus oligosporus*, and Sphaeriales, such as *Sordaria fimicola*, lack sucrase and are thus unable to utilize sucrose. But these are by far in the minority. Many of those which can utilize sucrose preferentially use the glucose moiety before fructose. Cellobiose, one of the hydrolytic products of cellulose is also widely used.

Of the polysaccharides, starch, the principal reserve of seed plants, is a good source for most fungi. It is often a better source for growth than is glucose. The utilization of a slowly hydrolyzable compound, such as starch, is often accompanied by the production and accumulation of smaller quantities of toxic by-products, such as acids, than is the utilization of the equivalent quantity of monosaccharides. The accumulation of such by-products in the immediate environment of the hyphae may eventually inhibit further growth. The very widespread ability of fungi to use such carbohydrates is reflected in the composition of standard culture media. The three commonest media employed are Czapek–Dox solution or agar, a basic mineral medium containing 3% sucrose, 2% malt extract agar containing maltose, dextrose (D-glucose) and dextrin (Table 1) and potato dextrose agar containing potato starch and 2% dextrose (DEVERALL, 1969).

Cellulose, the most abundant organic material on earth, cannot be utilized by all fungi but is slowly hydrolyzed by many. In order to degrade cellulose, fungi must again be able to produce the necessary hydrolytic enzymes. There are thought to be at least two different enzymes involved in its degradation since some fungi can hydrolyze modified cellulose, such as carboxymethylcellulose, but not native cellulose, such as cotton (Fig. 1–1). The exact mode of action of the first of these, called C_1, is not known but it produces linear chains. The second enzyme, Cx, which may be regarded as a 1-4,β-glucanase, catalyzes the hydrolytic cleavage of these to cellobiose. This enzyme also acts on modified cellulose. The cellobiose is then hydrolyzed to glucose by a β-glucosidase. Cellulases are usually only produced in the presence of cellulose and are not secreted in the presence of other readily available carbon sources. They are either bound to the surface of the hyphae and act on the surface with which the hyphae are in contact or they are secreted freely into the environment diffusing away from the hyphae. In the former case one sees zones of erosion around

Table 1 Constituents of Czapek–Dox agar and malt extract agar.

Czapek–Dox agar	
Sodium nitrate ($NaNO_3$)	3.0 g
Potassium phosphate (K_2HPO_4 or KH_2PO_4)	1.0 g
Magnesium sulphate ($MgSO_4 . 7H_2O$)	0.5 g
Potassium chloride (KCl)	0.5 g
Ferrous sulphate ($FeSO_4 . 7H_2O$)	0.01 g
Sucrose	30.0 g
Agar	15.0 g
Distilled water	1.0 l
If glass distilled water is used traces of zinc, copper and manganese must be added. Sterilize at 121°C for 15 min.	
Malt extract agar	
Malt extract	20.0 g
Agar	15.0 g
Distilled water	1.0 l

Unamended proprietary malt extract is suitable. Dissolve malt extract in warm water, add agar and steam until dissolved. Sterilize at 121°C for 15 min.

hyphae. The latter can be demonstrated by the clearing of finely powdered cellulose dispersed in agar in a Petri dish (Table 2, p. 16). Many cellulolytic fungi will clear the agar in front of the growing colony margin.

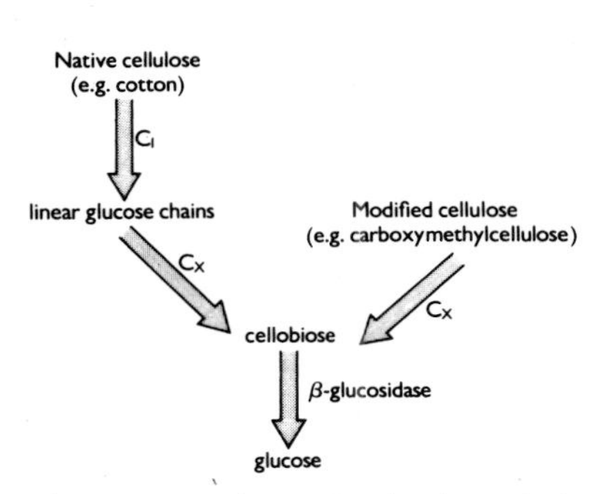

Fig. 1–1 A scheme for the degradation of cellulose.

The list of carbon compounds reported to be utilized for growth by fungi is almost infinite and includes, besides many other carbohydrates, amino acids, organic acids, sugar alcohols, lipids and alkaloids. Amino acids, in addition to providing a nitrogen source, may also serve as a sole carbon source but generally under such conditions ammonia accumulates and raises the pH to levels unfavourable for growth. Of the sugar alcohols, mannitol can be utilized by a much wider variety of fungi than can any of the others. For some it is equivalent to glucose as a carbon source. With trehalose and glycogen it is a widespread storage compound in fungi.

It is difficult to find a carbon-containing compound which some fungus cannot utilize. Certain fluorine containing plastics and a few detergents whose carbon cannot be used by micro-organisms may be the only examples.

1.3 Nitrogen sources

Fungi utilize inorganic or organic sources of nitrogen. Nitrates may be an excellent source for many fungi but inability to utilize them is common and ecologically important. Many Mucorales, such as *Mucor hiemalis* and *Rhizopus oligosporus*, cannot utilize nitrate and thus will not grow on culture media and natural substrata containing simple carbohydrates and nitrogen only as nitrate. This is also true of some Saprolegniales, many Blastocladiales and many wood-degrading Aphyllophorales and Agaricales, such as *Ganoderma lucidum* and *Pleurotus ostreatus*.

Few fungi are unable to utilize ammonia. Many reports of failure to utilize ammonia in culture may be due to inadequate buffering. The large fall in pH, following its uptake, is sufficient to stop growth. In mixtures of ammonia and nitrate, the former is preferentially absorbed. The very few fungi which are unable to utilize ammonia depend upon amino acids. These fungi include some of the aquatic Mastigomycetes, such as *Leptomitus lacteus*. Amino acids are thought to be used as such and built up directly to proteins rather than being degraded to ammonia first. Many fungi can use almost any amino acid as a sole nitrogen source. Asparagine is most often used in synthetic culture media. However, some fungi require one or two specific amino acids, others a whole range. Casein hydrolyzate is a good nitrogen source on which to culture such fastidious fungi.

Claims have been made that some soil fungi at least can fix atmospheric nitrogen. These have never been confirmed. In aerobic microorganisms nitrogen fixation is restricted to some blue-green algae and a few bacteria, such as *Azotobacter*.

1.4 Other requirements

Fungi also require a range of minerals, such as potassium and phosphorus, in quantity and others, such as manganese and zinc, in traces. These major and minor element requirements of fungi are very similar to those of other organisms.

In addition many fungi are heterotrophic for vitamins. The principal vitamins required are thiamin and biotin. Thiamin deficiency is particularly common in the Agaricales. The majority of species of *Boletus* and *Coprinus* are heterotrophic for thiamin. Some fungi are capable of synthesizing the thiazole moiety of the molecule and only need pyrimidine. The ability to synthesize pyrimidine but not thiazole is less common (SCOTT, 1969). Vitamins are required in exceedingly small quantities, in the order of 1–10 $\mu g\ l^{-1}$ or from 4–25 $ng\ mg^{-1}$ of mycelial dry mass as compared with about 5 $mg\ mg^{-1}$ mycelial dry mass of, for example, sucrose.

2 Cellulose Degradation and Wood Decay

2.1 Cellulose and the carbon balance

About one third of the organic matter produced by green plants is cellulose. It is an integral part of the primary and secondary cell wall. In fully mature woody tissues 40–60% of the dry mass is cellulose. In aspen wood, *Populus tremula*, for example, cellulose forms 44–48% of the dry mass. Cereal straws contain about 40% cellulose and seed hairs of cotton, some 95%. It is the major carbohydrate remaining in seed plants on their death and furthermore, it is not withdrawn from plant parts, such as leaves, when they are shed. Degradation of this cellulose is an indispensable process for the maintenance of the carbon balance in nature. Its degradation returns an estimated 85 billion tonnes (megagrammes) of carbon as carbon dioxide to the atmosphere each year. It has been stressed that if its degradation ceased while photosynthesis continued unabated, life on earth would stagnate for lack of atmospheric carbon dioxide in under 20 years.

2.2 Fungi and cellulose degradation

If we accept that the bulk crude source of organic carbon for heterotrophic saprophytes is wood and that to degrade wood cellulase production is a necessary prerequisite, the importance of fungi at once becomes obvious. The fungal hypha is well adapted to attack tough, bulky fibrous materials, such as wood. Fungal hyphae are septate or aseptate, tubular, uniseriate filaments, on average 1–15 μm in diameter. Growth in length occurs at the tip and is confined to this area so that in septate hyphae when a cell is cut off from the apex it is no longer capable of any significant increase in length. There is thus no increase in interseptal distances although there may be increases in diameter and wall thickness.

Fungal cell walls contain 80–90% polysaccharides with the remainder being protein and lipid. The two commonest building blocks of the former are D-glucose in glucans and 2-acetamide-2-deoxy-D-glucose in chitin. The majority of fungi with septate hyphae contain chitin and glucan in their walls. Chitin makes up 3–60% of the dry mass of the walls. The glucans are non-cellulosic and contain in the main 1–3, β- and 1–6, β-linked glucose. They are thus not degraded by cellulases. The Oomycete Mastigomycetes, such as the Saprolegniales and Peronosporales, have been traditionally regarded as possessing cellulosic hyphal walls with no chitin. In *Phytophthora*, for instance, glucans constitute about 90% of the dry mass of the walls and about one quarter of this has been reported to be cellulose, 1–4, β-linked glucose (Fig. 2–1). The remainder is a highly branched 1–3, β-linked glucan with 1–6, β-linked branches. The cellulose in the hyphal walls of such Oomycetes is poorly refractive and it seems more likely that it is not pure cellulose but a complex branched polymer with mixed 1–3, β- and 1–4, β-linkages. The other large group of aseptate fungi, the Zygomycetes, also appear to possess their own distinctive wall components. The two major components appear to be chitin and chitosan, which is like chitin but non-acetylated. D-mannose in mannans has been reported primarily from yeast cell walls and the combination of mannan and glucan appears to be characteristic of true yeasts and the yeast-like phases of other fungi.

Hyphal walls are usually two-layered with an inner layer of microfibrillar material, usually chitin but cellulose-like in Oomycetes, and an outer layer of amorphous glucan. They are significantly thinner at the apex but both layers are present to form the extendable wall. Behind the apex additional strength and rigidity are given by an increase in diameter, number and packing of the chitin microfibrils and an increase in thickness of the amorphous glucan (Fig. 2–2). Such addition must also mean that there is appreciable incorporation of wall materials behind the growing apex. Autoradiographs made of hyphae which have been fed with brief pulses of tritiated wall precursors have demonstrated that incorporation is highest in the apical 1 μm and falls off rapidly after the

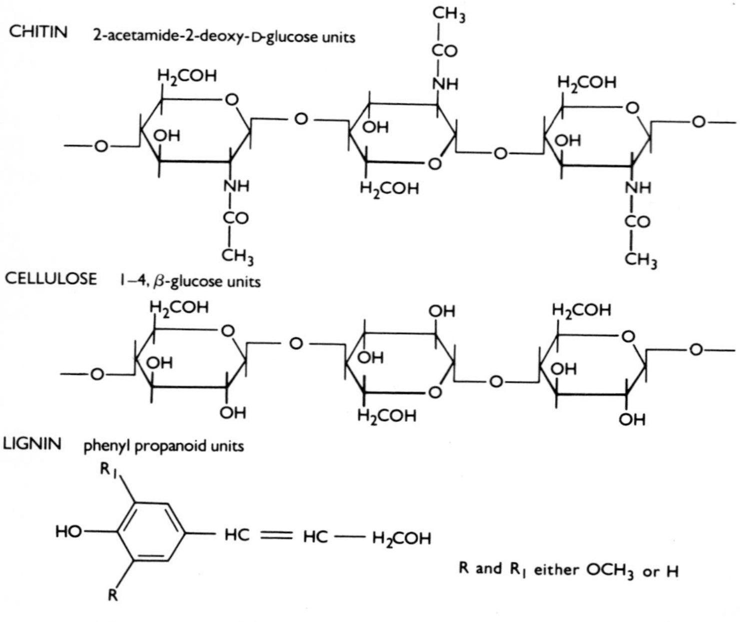

CHITIN 2-acetamide-2-deoxy-D-glucose units

CELLULOSE 1–4, β-glucose units

LIGNIN phenyl propanoid units

R and R_1 either OCH_3 or H

Fig. 2–1 Building units of chitin, cellulose and lignin.

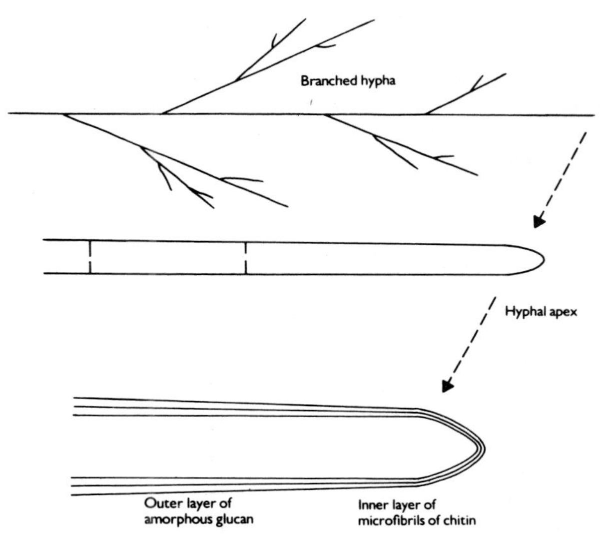

Fig. 2–2 Hyphal structure.

first 5 μm but that there is still appreciable incorporation from 5–75 μm behind the tip. This rigid nature of the wall behind the apex and the complex system of branching ensure that the older rearward part of the hypha is firmly anchored in the substratum and enable the hyphal tip to exert very considerable forward mechanical pressure as it extends under turgor. This, coupled with the production of extracellular enzymes which erode away the substratum, enables the hyphae to produce minute bore holes through cell walls. The much branched hyphal system can thus completely permeate even the hardest woody tissues. Thus cellulolytic fungi have an advantage over cellulolytic bacteria which can only accomplish breakdown by enzymic erosion at an already available free surface. They have no powers of rapid extension by mechanical penetration. They are of lesser importance than the fungi in the decomposition of wood under natural conditions, although they play a part in some heartwood rots and in rotting water-logged wood.

2.3 Sources of cellulases

The distribution of cellulolytic ability amongst the various classes of fungi is of considerable significance. Only a very few Mastigomycete and Zygomycete fungi, such as some members of the Chytridiales, produce cellulases. Cellulolytic ability is common, but not general in Ascomycetes and Fungi Imperfecti. The extent of the ability varies enormously from only slightly to markedly so. The most vigorous cellulolytic fungi are found in the Hymenomycete and Gasteromycete Basidiomycetes such as the Agaricales, Aphyllophorales and Lycoperdales. Many of these fungi not only attack cellulose but also utilize lignin. These are the most conspicuous of fungi, the mushrooms, toadstools, polypores and puffballs, especially common in woodlands.

Those fungi which can only utilize sugars and carbon compounds simpler than cellulose are often called 'sugar fungi'. These can be considered in two ecological categories, primary and secondary. The primary sugar fungi utilize any simple carbon compound initially available in the substratum. These rapidly colonize substrata rich in such soluble nutrients and disappear as these very accessible fractions are depleted. The secondary sugar fungi share the hydrolytic products of cellulose by growing in close association with the hyphae of cellulolytic fungi and absorbing some of the sugars as they are produced by the extracellular cellulases. The majority of the secondary sugar fungi produce β-glucosidases, so can utilize any cellobiose produced, and somewhat fewer produce the C_x enzyme. Some fungi may belong to both categories of sugar fungi and in using such categories it should always be borne in mind that other nutritional factors, such as available nitrogen sources, may be equally important as the carbon source in determining the ecological niche of these fungi.

Cellulolytic bacteria are more active in anaerobic or near anaerobic environments especially where there is a very large free surface area to volume ratio in their substrata, such as is found in the food material in the rumen of ruminant herbivores. The habit of 'chewing the cud' by the ruminant fragments the cellulose and presents a much larger surface area on which the bacterial cellulases can act.

Amongst animals, some insects, molluscs, a few Crustacea and Protozoa are said to produce cellulases. The best known cellulase producing animal is the snail, *Helix pomatia*, but although it is a voracious herbivore it rarely feeds on bulk cellulose, such as wood. Amongst arthropods, the silver fish (*Ctenolepisma lineata*) and the larvae of the death watch beetle (*Xestobium rufovillosum*) have been shown to produce cellulases in their intestines. This is also true of the earthworm, *Lumbricus terrestris*, but it is not always certain that these cellulases are not produced by bacteria as they are in the intestines of ruminants. The extent to which these, and other animals, utilize cellulose may be, in terms of the overall decomposition rate, secondary in importance to the comminution of the material. Their faecal pellets are usually a much more favourable habitat for fungi and bacteria than are the original food materials (section 3.10).

Compared with the relatively few animal sources of cellulases, there is a much more impressive list of cellulolytic fungi and it is these which are the principal decomposers of wood.

2.4 Wood decay

In addition to cellulose, wood contains 10–30% hemicelluloses and

20–30% lignin. Hemicelluloses consist of short heteropolymers of glucose, galactose, mannose, xylose, arabinose and certain uronic acids. These obviously require a number of enzymes to catalyse their hydrolysis. Little is known about these enzymes but they have been isolated from a number of fungi. Lignin, which is a three dimensional polymer of one or more phenyl propanoid monomers, such as coniferyl, sinapyl and coumaryl alcohol units, is attacked by relatively few fungi, primarily Basidiomycetes and a small number of Ascomycetes.

Three types of wood decay have been distinguished, brown rots, white rots and soft rots. In brown rots, caused by such fungi as *Piptoporus betulinus*, the carbohydrates, such as cellulose and hemicellulose, are attacked preferentially and there is little, if any, depletion of the lignin thus the wood becomes darker in colour. In white rots, caused by such fungi as *Coriolus versicolor*, both the carbohydrate and lignin components are attacked more or less simultaneously. Some fungi remove the lignin much more rapidly than the cellulose. The microfibrils of the latter are uncovered and the cellulose used later. In either case, the wood becomes much paler, often almost white, in colour. The fungi concerned in brown and white rots are mainly Basidiomycetes, especially the Hymenomycete Agaricales but more so the Aphyllophorales which are almost entirely confined to wood. Their hyphae ramify in the cell lumina and penetrate walls via pits or by producing bore holes somewhat wider than the hyphae. In white rots, there is a general progressive thinning of the secondary cell walls of the xylem outwards from the cell cavity, the enzymes acting in the near vicinity of the hyphae. Decomposition also occurs uniformly in the region of the attack. Whereas in brown rots there is no thinning of the walls. The enzymes diffuse away from the hyphae and act on the cellulose of the entire cell wall at some distance from the hyphae. This leaves a framework of lignin which maintains the general cell shape but with the structural fibrillar components removed, the walls have very little tensile strength and easily crumble. Decomposition also occurs in irregular patches in the attacked wood. This leads to the cubically cracked appearance of brown rotted wood.

The brown and white rots penetrate deep into the wood whereas soft rots are more conspicuous near the surface. In these latter rots carbohydrates are again principally utilized and they are caused by a number of Ascomycetes, such as species of *Chaetomium* and *Ceratocystis*, and some Fungi Imperfecti. Such rots occur in wood with an unusually high water content for example the fill of water cooling towers, underground mining timbers and river and marine timbers. The hyphae penetrate and grow within the secondary cell walls where they create enzymatically chains of cylindrical cavities, with pointed ends, more or less parallel with the microfibrils of the wall (Fig. 2–3).

Many actively cellulolytic fungi are restricted in their ability to utilize the cellulose in lignified plant cell walls by virtue of the intimate nature of the association between cellulose and lignin. A particularly good example

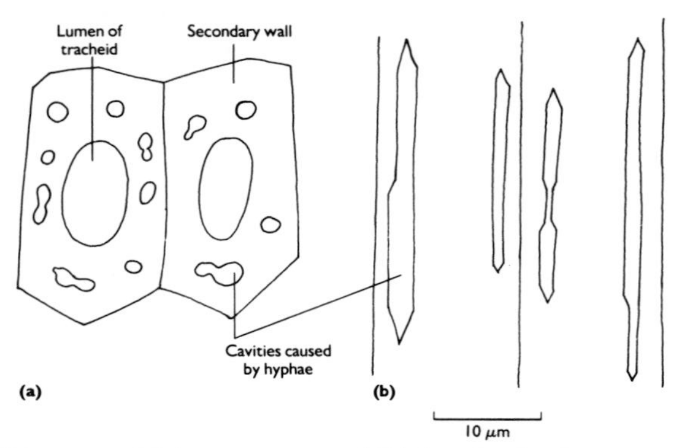

Fig. 2–3 Diagram of (**a**) transverse and (**b**) longitudinal sections of tracheids of *Pinus sylvestris* with soft rot cavities in the secondary walls. Note the characteristic pointed ends in (**b**).

is *Chaetomium globosum* which rapidly degrades cotton cellulose but only attacks wood of high water content and then merely produces cavities inside the cell walls as seen in soft rots. Lignin appears to act as a physical barrier that prevents the cellulases from reaching sufficient glycosidic bonds in the cellulose to permit large scale hydrolysis. Thus the accessibility of the cellulose to the degrading enzymes is a most important factor. This helps to explain the differences between brown and soft rots. The cellulases produced by the two types of rot are similar so that the differences are not associated with the different properties of the cellulases. It is suggested that the brown rots probably have other more effective enzymes which degrade substances, such as hemicelluloses which encrust the cellulose, and others which depolymerize lignin or at least disrupt its association with cellulose. This increases the accessibility of the cellulases so that they can diffuse freely away from the hyphae in contact with the cell walls. After such decay only the skeleton of predominantly lignin remains, which gives the decayed wood its brown colour. In soft rots other such enzymes are lacking and the activity of the cellulases is restricted to the exposed cellulose in the immediate vicinity of the growing hyphae. Similar increased accessibility can be achieved by fragmentation of the wood by, for example, ball milling. Increased hydrolysis of cellulose occurs in sawdust after ball milling and addition of cellulases, rather than addition prior to ball milling. The milling further breaks down the wood into much finer particles exposing a large surface of the cellulose free of its association with lignin. This may be the function of the chewing action of the termite, *Termes obesus*, which is a non-cellulolytic wood feeder. The fragmentation of the wood enables the cellulases produced by protozoans in its intestine to gain access to the cellulose. The termite is dependent upon these protozoans for its nourishment.

2.5 Resistance of wood to decay

Lignification of cell walls is thus an important factor that contributes to the natural resistance of sapwood and heartwood to fungal deterioration. Many other factors contribute to such resistance.

Wood degrading fungi have higher moisture requirements than most other fungal saprophytes. Whereas cotton is susceptible to fungal attack when it has a moisture content on a wet mass basis of more than 8% and cereal grains more than 13%, wood decay can only be initiated at moisture levels of about 26–32%, depending upon the wood. Dry timbers are thus not susceptible to fungal attack. The optimum moisture content of wood for fungal growth lies around 40% but again this varies with the wood and the fungus. Above the optimum moisture content fungal growth declines as less oxygen becomes available and the cell walls become saturated.

The low nitrogen content of wood also increases its resistance to decay. Woody tissues usually contain 0.03–0.10% nitrogen as compared with 1.0–5.0% in herbaceous tissues. Furthermore the nitrogen in wood is not uniformly distributed. It is highest in the cambium and pith and lowest just outside the pith. The carbon/nitrogen ratio of most woody tissues is thus high, in the order of 350–500:1, and may exceed 1000:1 in some heartwoods. For most fungi a culture medium with such a high carbon/nitrogen ratio would be nitrogen deficient and growth limiting. The wood-degrading fungi are adapted to this by possessing efficient mechanisms of assimilating, utilizing and conserving the meagre supply of nitrogen present. They use large amounts of carbohydrates to obtain sufficient nitrogen from wood in order to form their vegetative mycelium and reproductive structures. The vegetative mycelium of most fungi contains 3–6% nitrogen. This may fall to just above 1.0% in starvation. In some wood-degrading fungi, such as *Coriolus versicolor*, the level may fall lower to 0.2% in starvation. They also conserve nitrogen by the process of autolysis and re-use of the nitrogen content of their mycelium. Obviously this also contributes to their ability to bring about rapid decay despite the comparative deficiency of nitrogen. Although limited, the nitrogen content of wood is very variable and includes proteins, peptides, amino acids and amides, nucleic acids and inorganic nitrogen. All wood-degrading fungi can use ammonia, only a few use nitrate and they grow best on amino sources. They can thus use the majority of nitrogen sources in the wood.

The principal sources of decay resistance in wood are toxic substances deposited during the formation of the heartwood. Most attention has been paid to gymnosperm wood where many phenolic and other compounds have been extracted and shown to inhibit decay. These include terpenoids, tropolones, flavonoids and stilbenes (Fig. 2–4). Their mechanism of fungicidal action varies. Some of the stilbenes probably act as uncoupling agents which inhibit oxidative phosphorylation and thus decrease the main energy source of the fungi. In angiosperm wood tannins are more important and they would inhibit any fungal phenoloxidases.

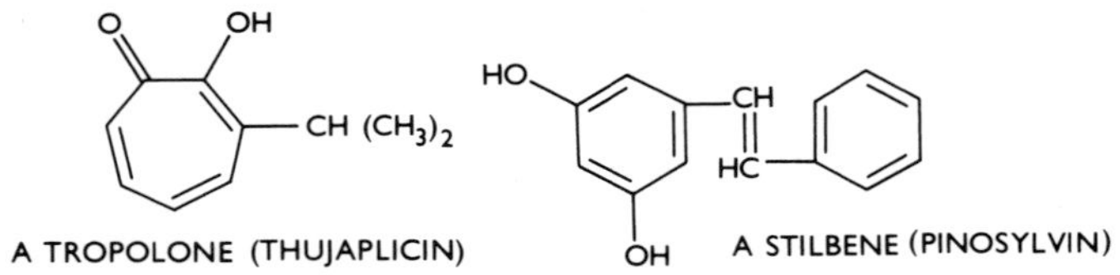

A TROPOLONE (THUJAPLICIN) A STILBENE (PINOSYLVIN)

Fig. 2–4 Structure of two toxic chemicals extracted from gymnosperm wood.

2.6 Dry rot

Any wood is completely immune to fungal attack as long as it remains continuously dry. This also applies to structural wood work in buildings. In houses, rotting begins in damp wood in embedded joists, behind skirting boards and in cellars. The most serious cause of wood decay in houses is *Serpula lacrymans*, the dry rot fungus. It establishes itself on a piece of damp wood and during active growth water is formed from the metabolism of the cellulose and excess is exuded in droplets, hence its specific name. The fungus once established and growing vigorously is thus able to provide itself with all the water it needs for growth and may then spread rapidly through dry timbers. It also produces mycelial strands which are an important feature of the fungus. The separate hyphae aggregate into complex strands which are more resistant to desiccation than are the individual hyphae so that they can grow out over exposed drier and non-nutrient surfaces such as bricks and plaster. It is this effective means of extension that makes it so potentially destructive once established. They also, and more important perhaps, form a means by which water and nutrients can be transported from a damp food base or substratum to be utilized in initiating an attack on dry timbers elsewhere in the house and thus its spread from room to room. Attacked wood gradually discolours brown as the cellulose is utilized and cracks into characteristic three dimensional cubes. Decayed wood is friable, light, dry, hence the name dry rot, and powders easily.

None of the timbers commonly used in this country are resistant to attack by *S. lacrymans* when placed under conditions favourable to the fungus, i.e. pockets with high moisture content and lack of ventilation to maintain high relative humidities. Such timbers can be treated with preservatives which confer immunity. Others such as wood of *Thuja plicata* (Canadian western red cedar), *Sequoia sempervirens* (red wood), *Cupressus nootkatensis* (yellow cedar) and *Tectona grandis* (teak) show a high resistance to attack. This resistance is conferred by the presence of such toxic chemicals as the thujaplicins (tropolones) formed in the heartwood.

Several other fungi frequently occur in houses but they only attack timber which is definitely wet and are to be found in any place where leakage of water is likely to occur. These fungi, such as *Coniophora cerebella*, the cellar fungus, require a high moisture content for growth and unlike *S. lacrymans* do not possess well-developed mycelial strands. They are very sensitive to drying and thus their growth can easily be checked by drying the timber.

Wood indefinitely submerged in water will also undergo slow fungal decay unless conditions are anaerobic. Some of the timber piles of the original London Bridge built in A.D. 1176 have been uncovered over the years and found to be sound and some of the oak piles of a Roman harbour buried beneath modern Dover have retained considerable tensile strength. There are very few marine Basidiomycetes and in the sea the major wood-degrading fungi are Ascomycetes and Fungi Imperfecti. Blocks of beech and other woods submerged in the sea soon became colonized by such fungi which produce, in addition to a surface rot, progressive degradation of the inner wood.

2.7 Blue stain of softwood

Another group of wood-inhabiting fungi are those which stain coniferous wood, especially the sapwood. Blue stains are caused by pigmented hyphae growing in the wood, whereas other stains such as brown stains, are caused by chromogenic substances actually excreted by the hyphae into the wood. At least one fungal stain has been used commercially. The mycelium of the Ascomycete *Chlorosplenium aeruginascens* permeates dead wood of oak and beech on the forest floor and colours it a brilliant green. Such 'green oak' is used for inlay and decoratively as Tunbridge ware.

Blue-stain fungi are non-cellulolytic. They utilize the more readily assimilable carbon compounds such as starch and sugars, which occur in the inner bark, sapwood and particularly the parenchyma of the vascular rays of freshly killed wood. They do no structural damage to the wood but merely discolour it. Blue staining is thus not the first stage of a form of rot, but it does indicate that the wood has been kept moist and exposed to conditions favourable to the development of decay fungi. Although blue stain has no effect on the structural properties of the timber, it is responsible for large financial losses to the producer. The mere discolouration of the sapwood makes architects disinclined to use it and it is less acceptable to manufacturers of packing cases and paper.

The term blue stain is applied widely to any grey or bluish discolouration of timber. Several types can be recognized. Deep blue stain is caused by fungi able to invade and grow rapidly through coniferous sapwood. Surface blue stain occurs most frequently on sawn timbers and on any wood exposed to rain and is caused by the surface growth of common airborne fungi with dark hyphae or spores, such as *Cladosporium herbarum*. Such staining is easily removed during planing treatment.

In vigorously growing pine trees the moisture content of the sapwood is too high to permit growth of blue-stain fungi. The low available oxygen supply again appears to be the major limiting factor in the growth of these fungi in wood with a high moisture content. In nature they colonize standing pine trees which have been killed either by *Fomes* root-rot or some other disease or by suppression but are more common on felled pine logs. These will all support growth of blue-stain fungi unless the moisture content of the sapwood falls below 27%.

The fungi concerned are Ascomycetes of the genus *Ceratocystis* and related Fungi Imperfecti. They produce sticky ascospores and conidia which are insect dispersed or distributed locally by rain splash. Their colonization is usually associated with the attack of bark beetles. Species of *Hylastes*, *Myelophilus* and others introduce spores in making their brood chambers at the interface of sapwood and bark. The spores germinate and grow radially and longitudinally in the sapwood forming wedges of blue-stained timber and then produce their sticky spores projecting into the brood chambers. These adhere to the young beetles as they emerge and are dispersed to other logs as they in turn make brood chambers.

After some 6–12 months in the sapwood they are replaced by *Trichoderma* sp. and wood-degrading Basidiomycetes such as *Peniophora gigantea* and *Stereum sanguinolentum*. *Trichoderma* sp. are common green moulds, the majority of which in addition to being cellulolytic, have marked antibiotic activity. The blue-stain fungi are very sensitive to these. By this time they have used most of the simple carbohydrates and conditions have become more conducive to the growth of cellulose- and lignin-degrading Basidiomycetes.

Blue-stain fungi can be readily isolated into culture where most will grow rapidly on 3% malt extract agar and soon produce either conidia embedded in a mucilaginous drop at the apex of conidiophores of a variety of types or perithecia with very long necks from which the ascospores exude in a droplet (Fig. 2–5). Such stalked spore drops in these and other fungi are usually associated with insect or rain splash dispersal.

A good source for these fungi are cut pine trunks, such as pit props or longer lengths, which have been left on the forest floor with their bark still attached. Staining will occur from after 2–6 months and the fungi can be isolated by inoculating Petri dishes of 3% malt extract agar with small fragments of stained wood. These can be removed from the wood by using either an increment borer or a small carpenter's drill. With either the bark should be removed to reveal the stained areas. Bark beetle activity will be obvious in such areas. The surface of the wood should be flamed and the tools sterilized by flaming after dipping in alcohol. The borer will remove a plug of wood which can be placed in a sterile container and sliced with a scalpel before plating out. With a drill, the drillings may be plated out. Some of the inoculations will undoubtedly

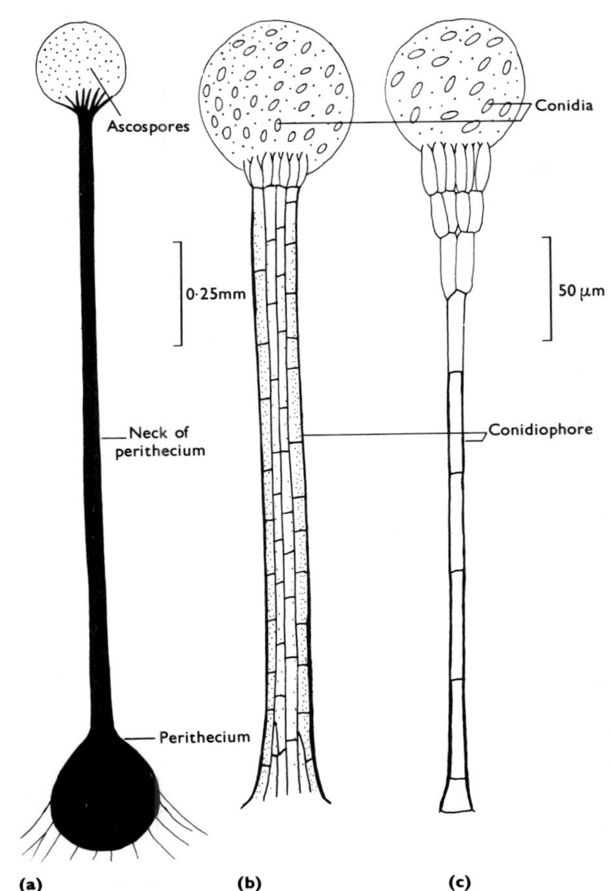

Fig. 2–5 Stalked spore drops of blue-stain fungi. (**a**) *Ceratocystis*, (**b**) *Graphium* and (**c**) *Leptographium*.

produce colonies of *Trichoderma* sp. which will be easily recognizable by their rapid growth rate and the green-coloured conidial masses which soon develop. Similar conidial masses can often be seen under the bark and at the cut ends of the logs under moist conditions.

Similar methods can be used to isolate wood-decay fungi. They, and cellulolytic fungi in general, will grow well on standard culture media (Table 1, p. 3), but attempts to isolate these from decaying wood often fail because on such media they may be overgrown by the more vigorous sugar fungi which may also be present. This may, to a certain extent, be overcome by the use of selective media.

The addition of 0.006% *o*-phenyl phenol to culture media suppresses fungal growth but the Basidiomycetes are less sensitive to it than other groups. If infected wood chips or sections are plated out onto this, colonies of Basidiomycetes slowly develop and can be sub-cultured to normal media such as 3% malt extract agar. Their identification in culture is usually difficult as basidiocarps only rarely develop and if they do they are not always typical.

Cellulose agar (Table 2) is another useful selective medium. Finely powdered cellulose forms the major carbon source and when poured into a Petri dish, it sets with the cellulose particles evenly distributed and is thus white and opaque. If decaying wood chips, litter fragments or other decaying cellulosic materials are plated out onto the agar or suspended in the agar before it sets, cellulolytic fungi will grow out from these and become obvious by gradually clearing the medium as they hydrolyze the cellulose. The medium does not fully inhibit the growth of sugar fungi but their growth is usually sparse so that they do not over-run the often rather slower growing cellulolytic species. Many Ascomycetes and Fungi Imperfecti sporulate more readily and more abundantly on this medium.

Table 2 Constituents of cellulose agar.

Ammonium sulphate ($(NH_4)_2SO_4$)	0.5 g
Potassium dihydrogen phosphate (KH_2PO_4)	1.0 g
Potassium chloride (KCl)	0.5 g
Magnesium sulphate ($MgSO_4 . 7H_2O$)	0.2 g
Calcium chloride ($CaCl_2$)	0.1 g
L-Asparagine	0.5 g
Yeast extract	0.5 g
Ball-milled cellulose	10.0 g
Agar	20.0 g
Distilled water	1.0 l

Standard grade cellulose powder is ball-milled for 72 h. The powder is prepared as a 4% suspension in water and is added to the other constituents after they have been mixed and steamed to melt the agar. Sterilize at 121°C for 15 min.

Sawdust is a useful medium for distinguishing between brown and white rots. About 20 g of sawdust from hardwood or softwood depending on the source of the fungi should be filled into boiling tubes after thoroughly mixing with 1 g maize meal, 0.5 g bonemeal and 40–60 cm^3 water. This quantity will fill a boiling tube to about 5 cm from the top. It should be plugged with cotton wool and autoclaved at 121°C for 30 minutes. When cool, an agar inoculum of the fungus should be buried to a depth of about 2.5 cm in the sawdust mixture. After about 14 days brown rot fungi will gradually darken the sawdust while white rot fungi will bleach the sawdust after producing an initial brown colouration. It is advisable to incubate the tubes in the dark at room temperatures. Basidiocarps may develop in such tubes if they are brought into diffuse light and the sawdust mixture is kept moist by constant addition of sterile water.

3 Woodland Fungi

3.1 Litter

The accumulated leaf litter and other debris in woodlands provides the most suitable of all habitats for the growth of the larger fungi, especially toadstools, polypores, puffballs and their relatives. It may be regarded as a bulk source of waste cellulose, other carbohydrates and lignin. Leaves form the bulk (60–70%) of this litter. The remainder consists of 1–14% bark, 12–15% branches and 1–17% fruit. The mean value of leaf litter fall in deciduous woodlands of the northern hemisphere is 3800 kg 10^4 m^{-2} yr^{-1}. It has been estimated that sessile oak (*Quercus petraea*) woods produce on average 24×10^6 leaves 10^4 m^{-2} yr^{-1}. These contain 41 kg 10^4 m^{-2}, or about 1%, of nitrogen and 1960 kg 10^4 m^{-2}, or about 50%, of carbon. The decomposition of this litter is of extreme importance. The return of carbon as carbon dioxide to the atmosphere is essential for the continuation of photosynthesis. The release of minerals, so that they again become available for absorption by plant roots, is equally important. The circulation of nutrients has been discussed by JACKSON and RAW (1966). They emphasize that the stages through which the carbon in the litter passes before finally being converted to carbon dioxide may be very varied and complex. Many fungi, other than the larger fungi, take part in this decomposition. Micro-fungi colonize the leaves before these cellulose- and lignin-degrading Basidiomycetes. Much of the litter fall is eventually eaten by the woodland fauna. A substantial part of what they eat is returned as faecal pellets and still other fungi play a part in their decomposition.

All these utilize part of the total carbon and in their respiration return some to the atmosphere and build up some of the remainder into their own tissues. These latter are degraded in turn on their death. The more complex of the carbon compounds, whether present initially in the litter or built up as part of the decomposer organisms, are resistant to decay and tend to accumulate in the humification layer of the soil. These include lignin complexes, chitin of fungal and insect origin and keratin of animal origin. This chapter is concerned with the ecology of fungi which colonize some of the various substrata found in woodlands.

3.2 Habitats of woodland Basidiomycetes

Apart from the species lists compiled from annual fungus forays surprisingly little ecological work has been carried out on these larger fungi. It is evident, however, that each seed plant community has its own characteristic fungal flora. This is clear from the results of WILKINS, ELLIS and HARLEY (1937). They recorded the fungal flora of five different types of seed plant communities. These were oakwoods, beechwoods, grasslands, coniferous woods and ling heath. The total number of species encountered in each and the number of species restricted to each ('exclusives') of the communities are given below:

	Oak	*Beech*	*Grass*	*Conifer*	*Ling*
Total	391	381	117	154	125
Exclusives	118	116	38	23	19

The exclusives were not absolute as they have been found in other habitats than the five studied. Only 14 species were common to all five communities and 42 were found in each of four. Three hundred and fourteen species were found only in one (exclusives). These and the 162 found only in two communities contribute to the characteristic fungal flora of each community.

The species which are constantly found in any one type of community are on the whole quite different from those in any other community. The fungi may be characteristic because they form mycorrhizal associations with the dominant tree present, e.g. *Amanita muscaria*, the fly agaric, with birches and pines, and *Boletus elegans* with larches. Others may be restricted to the wood of certain trees, whereas others may be restricted to a particular leaf litter, i.e. leaf-litter as distinct from wood-inhabiting fungi.

The habitats of wood-inhabiting fungi vary from minute twigs, large or small branches to the most massive of tree trunks and stumps. Such substrata are very complex and only in the simplest of terms can be considered as bark, sapwood and heartwood. As a group these fungi exhibit all degrees of specificity. Some may be restricted to a particular host genus. A good example is *Piptoporus betulinus* which is specific to birches (*Betula*). Birches are often killed by the fungus and diseased trees often break off at a height of about 3 m. The fungus then continues to grow as a saprophyte and produces its large bracket-shaped basidiocarps on both the fallen and standing part of the trunk. It is thus a facultative parasite. Other fungi may be restricted to a small number of hosts. *Polyporus squamosus* is most common on elm (*Ulmus*) but is often found on sycamore (*Acer*) (see LANGE and HORA, 1965). It is a wound parasite which causes a white rot of the heartwood but again it persists as an active saprophyte on its dead host where, like *Piptoporus betulinus*, it produces its basidiocarps for a number of years on the dead wood. Unlike the latter which is essentially autumnal in reproduction, it produces its basidiocarps in the late spring and early summer. *Auricularia mesenterica* is also usually found on elms whereas *Auricularia auricula*, Jew's Ear, is characteristic of elders (*Sambucus*) and is rarer on other hosts such as beech (*Fagus*) and elm. Still others, such as *Coriolus versicolor* (Fig. 3–1) and *Stereum hirsutum* are much less discriminate and grow on a wide range of woods. The former is one of the commonest fungi on stumps of hardwoods where it produces a rapid white rot but, like *Stereum*, it is usually confined to the sapwood. Many fungi show a preference for coniferous wood. *Tricholomopsis rutilans* and *Paxillus atrotomentosus* (Fig. 3–2b) are characteristic of coniferous stumps, *Polystictus abietinus* of coniferous branches and *Auriscalpium vulgare* of pine cones. These fungi are all Basidiomycetes but similar examples can be found in the Ascomycetes.

Fig. 3–1 Basidiocarps of *Coriolus versicolor* on a plum stump. The stump has been cut to show the white rot in the sapwood ($\times \frac{1}{3}$).

For example, *Daldinia concentrica* (Fig. 3–2a) is characteristic of ash (*Fraxinus*) but is occasionally found on other wood, whereas *Xylaria polymorpha* is common on a wide variety of hardwoods. A great variety of smaller Ascomycetes appear with these and the Basidiomycetes but they again are confined to the bark and sapwood. They also may be host specific or host restricted.

Each tree species thus has, within limits, its own particular fungal flora. As suggested, many of the fungi concerned may enter as parasites at wounds above ground or along roots below ground and they persist as

Fig. 3–2 (**a**) A perithecial stroma of *Daldinia concentrica* growing on a branch of ash ($\times \frac{1}{2}$). (**b**) Basidiocarps of *Paxillus atrotomentosus* growing on an old stump of pine ($\times \frac{1}{7}$). (**c**) Basidiocarps of *Hypholoma fasciculare* growing around an overgrown and well-rotted stump of pine ($\times \frac{1}{4}$).

active saprophytes after death. They have the competitive advantage over saprophytes in being established first on their particular substratum. As parasites they may be able to parasitize only one or a few species of plants. This would account for some of the specificity noted. The fungi growing in the heartwood of a particular tree are often different to those utilizing the sapwood. Here it may well be that the different naturally occurring tannins, terpenoids, etc., present in the different heartwoods further help to determine specificity. Only fungi which can tolerate or degrade these can become established.

Many other saprophytic fungi are confined to the decaying leaves on the woodland floor. Some of these will grow on the leaf litter of all kinds of woodland. These include such common toadstools as *Collybia maculata*, *C. peronata*, *Clitocybe nebularis* and *Laccaria laccata*. They may, however, be more common in particular types of leaf litter. The first, for example, is found predominantly in pine but also in beech litter. Others are more restricted. *Marasmius androsaceus*, for example, is found on pine needles but also occurs on heather, whereas *M. wynnei* occurs on beech leaves. Such fungi are usually neglected in ecological studies of woodlands because of the difficulties of identification except when basidiocarps are present but they are no less a part of the community than are the seed plants.

3.3 Environmental effects on the fungi

The distribution of fungi is primarily controlled by the distribution of their substratum or host on which they are growing. Climatic and edaphic factors have less effect upon their distribution than they have upon that of seed plants. Hence we speak of fungi as 'common in beechwoods' or 'common on oak'. However, it is a matter of observation that environmental conditions, in particular rainfall and temperature, have a very marked effect on both the time of appearance and the number of basidiocarps produced.

WILKINS and HARRIS (1946) recorded the number of fungi, the rainfall, the temperature and the effect of leaf canopy on the last two in a beechwood and in a pinewood. In both woods no basidiocarps were found from January to May, but from the start of the season in June there was a steady increase in numbers which reached a peak in October and then decreased until December.

The amount of rainfall determines the moisture content of the litter. Provided that this is above a certain level and and that the temperature is within a certain range, basidiocarps will appear after a certain time. Many species appear relatively indifferent to temperatures within limits. Usually the minimum temperature has more influence than the maximum. A low minimum directly retards mycelial growth in the litter, whereas a high maximum is indirectly effective as being associated with a low moisture content. It is very probable that given a favourable moisture content of the litter and a favourable temperature, basidiocarps could be produced at any time of the year. Thus the rhythmic periodicity of their appearance is more a function of conditions than of time. They estimated that in a beechwood basidiocarps could be produced when the moisture content of the litter was above 50% and when the mean of minimum surface temperatures is above 4°C. These two only coincided in the wood they investigated from the beginning of June to mid-August and again from the beginning of September to mid-October. These two periods were the only times that basidiocarps were produced in any quantity. The exact extent of the reproductive period in any one year obviously depends very much on the weather. The mid-August break noted by Wilkins and Harris is dependent on rain or lack of it, the effect of leaf canopy on its penetration and the high temperature causing its evaporation. In a cool wet August one would not expect such a break.

The length of time during which favourable conditions must operate before fungi reproduce varies with the species but there must obviously be a lag. In general, the larger the basidiocarp, the longer the lag. The smaller species appear first and only if the favourable period continues for some time will larger species appear. Most of these fungi have a perennial mycelium which persists in the litter throughout the year. It becomes inactive when conditions are too cold or too dry but revives under favourable conditions. Basidiocarps develop only when the mycelium has grown sufficiently to amass the necessary reserves from the litter.

The leaf canopy has a marked modifying effect. It reduces the amount of rain reaching the litter. In a beechwood in full leaf canopy only about 55% of the rainfall reaches the litter. This is disadvantageous for growth of the fungi but is more than compensated for by the modification of the temperature. During May to September outside the beechwood studied the average maximum soil surface temperature was as high as 45°C, whereas in the wood it rarely exceeded 30°C. The average minimum surface temperature was also higher in the wood by about 5°C. Thus the detrimental effects of the heat of the summer sun and the cold of the early frosts are avoided. Indirectly the canopy increases the relative humidity by preventing insolation and by reducing air movements. On the whole it seems that woodlands offer a better habitat for fungi than open habitats such as grasslands. The modifying effect of the canopy enables the fungus reproductive season to start earlier and continue longer by maintaining more equable conditions. This applied equally to leaf-litter and wood-inhabiting fungi. Wood retains moisture better than the leaf-litter and the mycelium of the wood-inhabiting fungi is better protected within the wood from desiccating conditions.

3·4 Experimental ecology

Experimental ecological work on these larger fungi has been very limited. Further experiments along the lines of those initiated by HORA (1959) are required. He experimented with various mineral additions to small plots in plantations of Scots pine (*Pinus sylvestris*) growing on mor type peat with a sparse ground flora on acid gravel. He recorded over 40 Basidiomycete species in the plantations. The plots were treated with either hydrated lime, superphosphate, sulphate of ammonia, 'Growmore' or potassium nitrate in the winter months. In the following autumn three species occurred in greater profusion than the others. Two of these were common in coniferous woods, *Lactarius rufus* and *Paxillus involutus*. The peak reproductive season for these was mid-September. 'Growmore', ammonium sulphate and superphosphate were remarkably effective in stimulating basidiocarp formation. For example, the untreated control plots yielded a maximum of 50 basidiocarps per week of *L. rufus* in each 10 × 20 m plot whereas addition of 'Growmore' yielded over 200. Lime had a depressing effect on both. This is understandable in that, by comparison with seed plants, they can both be classified as calcifuge species. The potassium nitrate had only a small stimulatory effect on *L. rufus* and was inhibitory to *P. involutus*. Acid litter and soils do not normally contain nitrates and many of the fungi which grow in these cannot utilize nitrate nitrogen but utilize ammonia nitrogen, hence the effect of 'Growmore' and ammonium sulphate. The inavailability of nitrogen is probably one of the major limiting factors for growth of fungi in woodland litter. Its addition makes more of the carbon in the litter more rapidly available. But it is not the only factor. Similar but less marked stimulation by superphosphate indicates that the inavailability of other minerals, especially phosphates, may also be limiting growth.

The third species, *Myxomphalia maura*, only occurred on the limed plots and was completely absent outside these. Its normal habitat is bonfire sites on acid soils in coniferous woods. It is not found in woods on limestone soils so it is not a calcicole. It appears to be encouraged by the sudden addition of alkali to acid soils. Three plots were limed in three different plantations with a single application of 203 g m^{-2} of hydrated lime during three consecutive winters. In the first season after liming a total of 18 basidiocarps was recorded in the three limed plots, in the second over 18 000 and in the third over 40 000 with a maximum weekly count of nearly 7800 basidiocarps. This poses a number of intriguing nutritional problems. For example, why does it not occur in litter above limestone soils after burning? Similar experiments to these in a variety of woodlands would no doubt pose further problems but could also yield valuable information regarding some of the factors influencing the distribution of particular fungi. Such information at present is extremely scarce.

3.5 Pyrophilous fungi

Several other Basidiomycetes, such as *Pholiota carbonaria* and *Collybia atrata*, are like *M. maura* in that they occur in nature only on burnt ground. This is also the habitat of about 50 British apothecial Ascomycetes. The best known genus of these pyrophilous fungi is *Pyronema*, whose name implies that it inhabits burnt places. It is a very common colonizer of bonfire sites and is often conspicuous because of the multitude of small confluent pink apothecia which it produces. It is also common on steam-sterilized soil in greenhouses.

The carbon and nitrogen requirements of these fungi are no different from other fungi but a number of important ecological factors obviously operate in confining their growth and reproduction to burnt ground. Fresh bonfire ash is very alkaline with a pH of about 8.0 to 9.0. The heat of the bonfire, together with the increased alkalinity associated with the ash, may reduce or eliminate completely the normal fungal population. Thus competition is greatly reduced. Recolonization of the ash is limited to those fungi which can tolerate such alkalinity. Many of the pioneer colonizers have higher pH optima than do other fungi. Their growth rates are also increased with increased alkalinity. Many also require alkaline conditions for the development of their ascocarps, apothecia. They thus prefer alkaline conditions unlike the majority of soil fungi which prefer slightly acid conditions. For example, the ascospores of *Pyronema domesticum* germinate best at pH 7.0–8.0, whereas the sporangiospores of *Mucor hiemalis*, a common soil fungus, germinate best at pH 4.5. *Pyronema* grows over a wide range of pH but grows fastest at pH 8.0 and the optimum pH for the production of apothecia is 8.3.

After dispersal the ascospores of these fungi remain dormant and often a heat trigger, such as a short exposure to temperatures around 50°C, is required to induce their germination. At bonfire sites spores are present on the surface or in the subsurface layers of the soil. At a particular distance from the centre of a bonfire the spores receive the correct heat treatment and germinate, inside the distance they are overheated and killed and outside remain dormant. The activated spores germinate rapidly and exploit the ash by virtue of their rapid growth rate. This gives these fungi first access to the charred wood and other plant remains.

3.6 Micro-fungi of leaf litter

Before Basidiomycetes appear in the leaf litter a large variety of micro-fungi have already colonized the leaves. Colonization begins on the tree. Once unfolded, the leaves act as spore traps for any airborne spores. An immense variety of spores can be found on leaf surfaces, or any plant surfaces. Their presence merely reflects their relative abundance in the atmosphere and the efficiency of the leaves as spore traps. This can be demonstrated by making leaf surface impressions by covering the leaf surface with nail varnish or spraying with a 1 : 1 solution of clear cellulose dope and acetone (or commercial thinners) and peeling off the resultant films when dry. Fungal spores become embedded in or adhere to such films and can be stained in lactophenol containing 0.1% fuchsin. Similar films can be prepared by appressing 'Sellotape', stripping off and staining. The catch will be comparable to that of the trace from an impaction spore trap, such as the Hirst or Burkard volumetric spore trap or a rotorod sampler (GREGORY, 1973).

Some of the spores germinate immediately and grow on the leaf surface and form the phylloplane or leaf-surface fungi. The majority of these are yeast-like forms and the commonest is *Sporobolomyces roseus*, one of the 'shadow yeasts'. Its presence can be demonstrated by suspending leaves from the underside of the lid of a Petri dish with vaseline for about 12 h over 2% malt extract agar. It is a Basidiomycete yeast and in such a humid atmosphere each cell produces a single basidiospore which is discharged and falls onto the agar below. Each forms a small budding colony, pink in colour and visible to the naked eye after 2–3 days of incubation. These colonies form a mirror image of the distribution of the cells on the leaf surface. These, and others, grow as saprophytes on the leaf surface living on simple organic substances such as sugars, amino acids and inorganic ions, which are exuded or diffuse out of the leaf. They persist as surface inhabitants until after leaf-fall.

With leaf senescence any facultative parasites within the leaves may persist and spread. Spores of leaf saprophytes present on the leaf surfaces germinate and rapidly colonize. The majority of these are Ascomycetes and Fungi Imperfecti, the commonest of which are *Cladosporium* sp., *Epicoccum purpurascens*, *Aureobasidium pullulans* and *Alternaria alternata* (Fig. 3–3). These are all very common airborne fungi. *Cladosporium* forms dark olive green sporing colonies on all sorts of plant debris and produces dry conidia which are easily dislodged by a breeze (Fig. 3–4). It is by far the most frequent of these fungi. This is reflected in the air-spora. In the summer months air contains from 800–12 000 spores m^{-3} of *Cladosporium* and they form 60–80% of the total air-spora in dry weather. They can be easily isolated from the atmosphere by exposing Petri dishes of nutrient

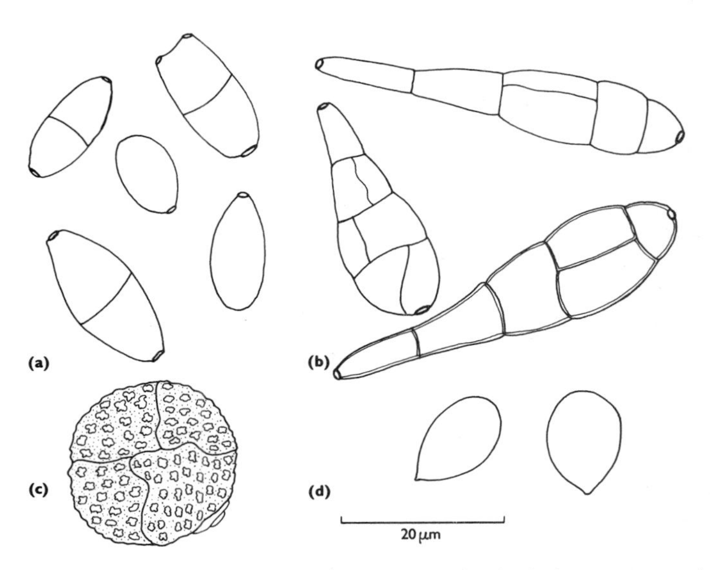

Fig. 3–3 Detached conidia of some primary leaf-inhabiting saprophytes. (**a**) *Cladosporium,* (**b**) *Alternaria,* (**c**) *Epicoccum* and (**d**) *Botrytis*.

agar to the atmosphere for 5 to 10 minutes on a horizontal surface at about 1 m above ground level. Colonies will appear after 2–3 days of incubation at 20–25°C and will be recognizable after 5–7 days by their olive green colour.

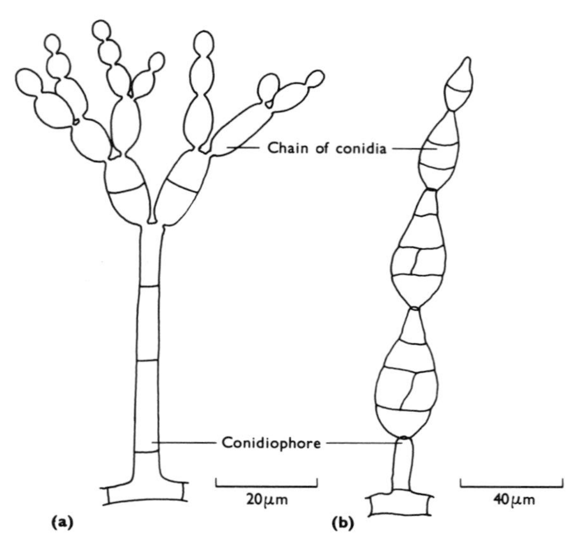

Fig. 3–4 Conidia and conidiophore of (**a**) *Cladosporium* and (**b**) *Alternaria.*

Most of these fungi utilize simple carbon compounds, such as sugars and starch, but many have the ability, in culture at least, to utilize cellulose and are thus not strictly sugar fungi. With leaf-fall they increase in frequency but are eventually replaced by a larger variety of other saprophytic Ascomycetes and Fungi Imperfecti. It is these that are eventually joined by Basidiomycetes in the litter. The leaves become progressively fragmented by the activity of the fungi and other micro-organisms present in particular the micro-fauna. With progressive fragmentation they become incorporated into the soil and colonized by true soil-inhabiting fungi, such as *Mortierella*, *Mucor*, *Penicillium* and *Trichoderma* sp. (Fig. 3–5). This general outline of the fungal succession on tree leaf litter is depicted in the schema below (Fig. 3–6).

3·7 Decomposition of pine needles

The time period between leaf-fall and its final decomposition varies enormously. In Cool Temperate pine forests it may be ten years, in ash and sycamore woods under one year, in Tropical forests mere weeks. Pine needles are very durable and only decay slowly, particularly above acid soils, so that there is an accumulation of a considerable bulk of leaf litter layered in different ages at different stages of decay. The needles are shed in

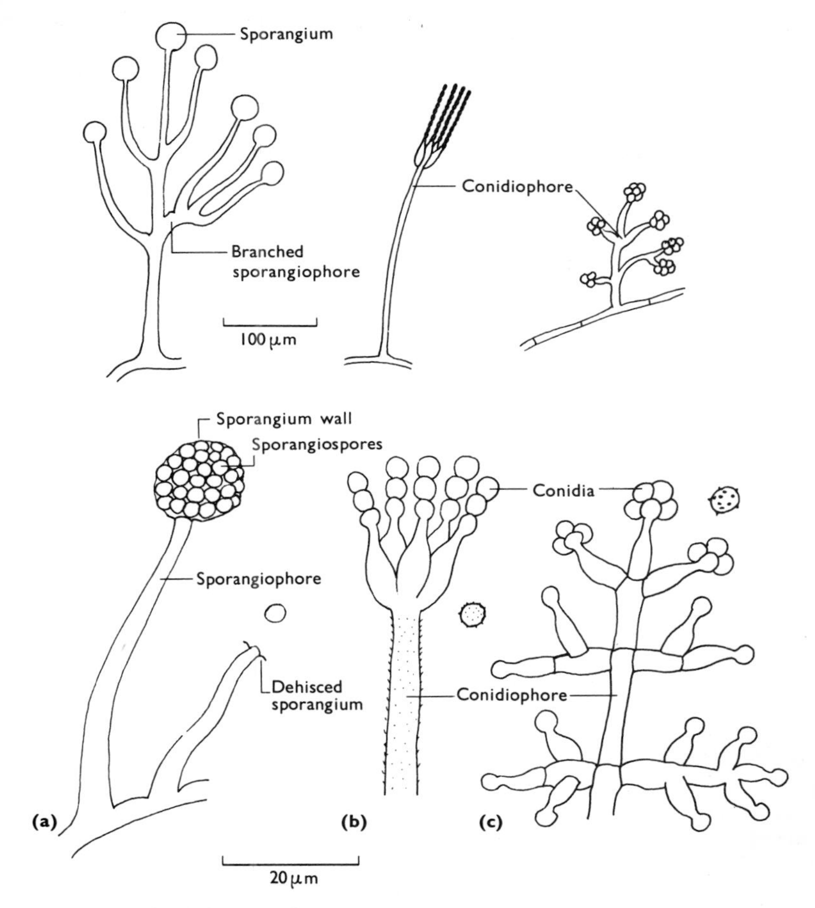

Fig. 3–5 Soil-inhabiting fungi. (**a**) Sporangia and sporangiophores of *Mortierella*, (**b**) conidia and conidiophores of *Penicillium* and (**c**) conidia and conidiophores of *Trichoderma*.

August and September and form part of the litter layer and they remain there for about 6 months before being incorporated into the fermentation layer. The litter layer is thus composed of recently fallen needles, light brown to buff in colour, and others, somewhat darker in colour, which have fallen earlier. They all have high tensile strength, a relatively low moisture content and lie in a loose uncompacted layer on the surface. In this layer the leaves are very susceptible to drying out and conditions are least favourable for fungal growth. The most conspicuous of the fungi is *Lophodermium pinastri*, which forms black, elliptical ascocarps opening by a slit. It is a facultative parasite which persists as an active saprophyte for

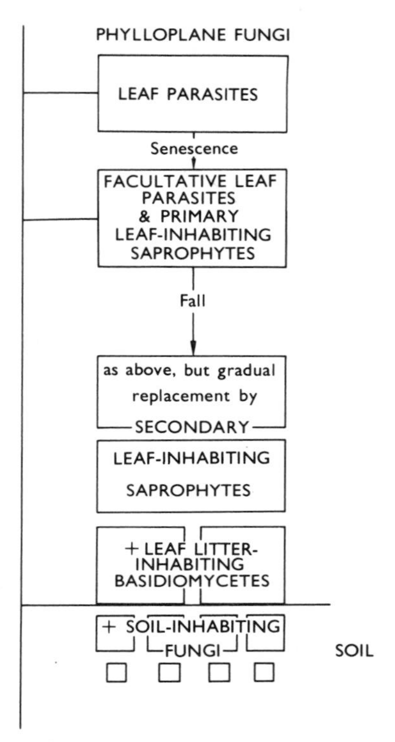

Fig. 3–6 General outline of the fungal succession on tree leaves.

over a year after leaf-fall. It is accompanied by a number of smaller saprophytic Ascomycetes and Fungi Imperfecti.

The fermentation layer can be divided into two. In the upper layer the needles are grey but gradually darken with depth and have softened tissues with low tensile strength and a high moisture content. They remain in this layer for about two years. It is in this layer that maximum development of Fungi Imperfecti occurs. The interiors of the needles become intensively attacked and the exteriors become covered by an extensive fine network of dark hyphae. It is these which contribute to the darkening of the needles. In the lower layer the character of the needles again changes. They are greyish, fragmented and compressed. Mites eat their way through the needle fragments and enchytraeid worms, dipterous larvae and Collembola feed on the fungi present. Most needle fragments bear dark amorphous faecal masses of the micro-fauna in which fungal remains can be detected. The fungi most active in this layer are the leaf litter-inhabiting Basidiomycetes, such as *Lactarius rufus*, *Collybia maculata* and *Paxillus involutus*, which slowly decompose the needle fragments over a period of about seven years. It is from this layer that true soil fungi can be isolated.

Eventually the remains enter the humification layer which consists largely of faeces and the remains of both the micro-fauna and the fungi, the needles having undergone complete physical reduction. The humification layer is rich in chitin in the form of hyphal wall fragments and exoskeletons of insects or other chitinized remains of the micro-fauna. Some of the fungi present have the ability to break down this very resistant substratum and their activity may well represent one of the final stages of the mineralization of primary and secondary organic materials in the litter. In this layer conditions are very acid, amorphous humus accumulates and biological activity is at a very low level.

3.8 Chitinolytic fungi

The structure of chitin (Fig. 2–1) is like that of cellulose except that an acetyl amino group replaces the hydroxyl of the number two carbon atom. The necessity for its decomposition is self-evident. Its continued accumulation would seriously deplete available carbon and nitrogen sources. Several billion tonnes of chitin are produced each year by copepod crustacea alone.

The activity of chitinolytic fungi can be demonstrated by 'baiting' soil or humification layer litter with chitin. Chitin as it occurs in nature may be mixed with other organic materials sufficient in amount to provide the carbon source for fungi so that it cannot be claimed that a fungus utilizes chitin, i.e. is chitinolytic, merely on the evidence that it grows, for instance, on insect exoskeletons. Chitin powder can be obtained commercially but natural chitin is difficult to purify. Reasonably pure chitin can be prepared from *Sepia* shells. They are first treated with 10% HCl to remove carbonates. The acid is changed daily until carbon dioxide emission ceases. The shells are then deproteinized by treatment in excess 10% NaOH at 103°C for two six hour periods separated by a six hour period in 5% HCl. After washing and autoclaving, strips can be mounted on glass slides and placed in jars of soil or humification layer litter. After 3–4 weeks' incubation the strips are washed in sterile water to remove adherent soil and fungal propagules and placed on the surface of chitin-containing mineral agar (10 g l^{-1} chitin powder). Any fungi which grow out can be screened for chitinolytic activity by transferring to chitin strips in moist chambers or to chitin-containing agar. This is a useful enrichment method and has helped to reveal that species from several common genera occurring in the humification layer and in soils, such as *Mortierella*, *Penicillium* and *Trichoderma* (Fig. 3–5) are chitinolytic.

3.9 Mull and mor

In a pinewood each successive leaf-fall buries the previous one so that a stratified litter layer, as described, is produced. The animals in this litter are sufficiently small for them not to disturb this stratification to any extent. This type, with a substantial litter layer remaining relatively undisturbed, has been called mor. Mor formation is associated with leaves rich in phenolic compounds and of very low base status. The low pH tends not to support large earthworm and bacterial populations.

The extreme contrast is the mull type which usually occurs under more alkaline conditions. Incorporation of the litter into the soil is much more rapid and is facilitated by earthworms which cut up the leaves and remove them to their burrows. Here the more favourable moisture regime is conducive to decomposition. In this type it is not usually possible to distinguish a clear junction between litter and soil. Under mull conditions, oak and hazel leaves may be incorporated into the soil in 7–9 months whereas 9 years may be required to incorporate pine needles under mor conditions. All intermediate conditions exist. Beech leaves, with their higher tannin content than most deciduous tree leaves, may take 3–5 years before being finally decomposed.

Many factors must operate in influencing what determines the differential formation of mull and mor. It has been suggested that phenolic compounds in the leaves tan proteins which are abundant in the mesophyll. These tanned proteins are very resistant to decomposition and form a protective covering to the cellulose of the cell walls so that they cannot be degraded by fungal cellulases. The tanned proteins, in addition to resisting decomposition and masking cellulose, delay the decomposition of unprotected tissues by withholding normally available nitrogen.

3.10 Decomposition of faecal pellets of the micro-fauna

Leaves themselves are very complex substrata and during their decomposition produce a variety of equally complex secondary substrata such as faecal pellets and chitinous remains of the micro-fauna which feed on them. The pellets are a large potential source of seed plant nutrients and their decomposition is of some importance.

Much of the annual litter fall in woodlands is eventually eaten by the soil micro-fauna but usually more than 60% and often as much as 80–90% of that eaten is returned as faecal pellets. This includes all the lignin and varying proportions of the cellulose. The exact amount of the latter depends upon such factors as whether or not the animal, or its intestinal micro-flora, produces cellulases or not, the amount produced and the extent to which the material is fragmented thus enabling the enzymes access to it.

It has been claimed that a single larva of the terrestrial caddis fly (*Enicocyla pusilla*) can break down a single oak leaf into some 3000 faecal pellets. Only about 7% of the leaf material consumed was utilized so that 93% was returned to the litter as faecal pellets. Oak leaves contain 20% cellulose, 40% lignin and 25% non-cellulosic carbohydrates. The larva used just over half of the latter but no cellulose or lignin. By comparison the millepede (*Glomeris marginata*) when fed on hazel (*Corylus*) leaves again returned over 90% as faecal pellets and these included over 70% of the cellulose. The pellets contained more ammonia nitrogen and total nitrogen g g^{-1} than did the leaves. There was more than a seven-fold increase in ammonia content.

Thus the litter micro-fauna eat relatively large amounts of leaf litter of little nutritional value and excrete most of it unchanged chemically but nevertheless greatly fragmented and richer in nitrogen. The overall rate of decomposition of the leaf litter is usually increased when it is converted to faecal pellets. Many factors and micro-organisms are involved. The fragmentation by the micro-fauna presents a large surface area for extracellular enzymes to act upon. This particularly favours bacterial activity. The pellets have a better water holding capacity than the natural leaf litter. Leaf litter is more prone to drying out at least until compacted and is thus less favourable for rapid and continuous mycelial growth. The higher nitrogen content favours synthesis of microbial protein and thus increases microbial growth and the decomposition rate.

On such pellets there is a peak of microbial activity during the earlier stages of the decomposition. This peak is associated with the build up of large populations of rapidly growing micro-organisms such as bacteria and Zygomycete Mucorales utilizing the easily assimilable carbohydrates present. After the peak, bacterial numbers fall, while the amount of septate fungal hyphae increases slowly as the pellets age. After the Mucorales, a wide range of Fungi Imperfecti and Ascomycetes, such as *Chaetomium* sp., appear. Their appearance is coupled with the slow utilization of cellulose. Eventually Basidiomycetes may colonize. The normal fungal succession on leaves is disturbed when they are converted to pellets. The secondary succession which develops has an initial phase of Zygomycete Mucorales, not Ascomycetes and Fungi Imperfecti. A similar, but better developed, fungal succession is seen on herbivore dung. This succession is discussed in the next chapter.

3.11 Wood decay and mineral cycling

Wood decay is important in regulating the cycling of mineral nutrients in the woodland ecosystem and contributes to the processes involved in soil development there. Over the period of active fungal decay virtually all the important minerals, but in particular nitrogen and phosphorus, are immobilized in an organic form in the fungal hyphae or basidiocarps. Although the mineral content is low per unit volume, the sheer volume of decomposing wood means that it forms a very substantial part of the total minerals in the woodland ecosystem.

The incorporation of decaying wood into the soil is greatly facilitated by animals. As brown rot fungi, and white rot ones, exploit the wood, it softens and becomes friable and as such is much more attractive to animals, not only as a food source but as somewhere to live and breed. Initial colonization by wood-boring beetles provides ideal entry channels for a great variety of animals generally common in the litter and soil. As in leaf litter, these include mites, Collembola and Isopoda as well as enchytraeid and lumbricid worms. Many of these and in particular the mycetophilid dipterous larvae feed mainly on the mycelia of fungi. These all again accelerate the process of fragmentation and incorporation and, in addition, inoculate the wood with true soil-inhabiting fungi, by carrying in spores from the surrounding litter and soil. Zygomycete Mucorales appear for the first time on such substrata along with a variety of species of *Penicillium* and *Trichoderma*. These now have a very wide choice of substrata on which to grow. They may utilize the partly degraded wood, the dead hyphae of the decay fungi, the bodies of the dead animals or their faecal remains. Some may even live as commensals sharing the hydrolytic products of the decay fungi.

Comminution of the wood by animals leads to release of some minerals. The small particle size of the frass produced by the borers and faecal material means that some are more effectively leached. The wanderings of animals, after feeding, leads to some redistribution of minerals. As in the fungi, the biomass of the animals acts as a reservoir of mineral nutrients in a very much more concentrated form than that found in the wood itself. Export of such materials may occur as adults emerge from the wood, but this is only a redistribution to other parts of the ecosystem. These minerals are only made available when the animals die and their tissues are mineralized. Chitinolytic fungi play an important part here.

3.12 Decomposition of lignin and humus in the soil

Wood decay fungi causing white and brown rots are more often associated with relatively large masses of wood such as the dead tree trunk, an old stump or a fallen branch. This may only be because these contain enough energy resources for these fungi to amass sufficient reserves to produce their relatively conspicuous basidiocarps. Vast quantities of lignin are incorporated into the soil in the vascular network of leaves, fine roots and so on. These are very different substrata for fungi and other micro-organisms and they are in a vastly different environment. The substratum is richer in terms of associated readily available carbon and nitrogen sources, other tissues being present in addition to the xylem. It is also presented to a much more varied population of micro-organisms. As such it would support a more diverse micro-flora and any lignin decomposers would find competition for not necessarily lignin but other more easily assimilable products which are necessary for establishment and be exposed to antagonism by others. This situation is markedly different to the decaying tree trunk with its one or few decomposer fungi in isolation.

Large quantities of lignin may also be introduced into the soil in the form of organic residues from wood decay. Whereas white rot fungi will eventually degrade lignin completely to carbon dioxide and water, brown rot fungi leave the lignin in wood virtually intact and this residual lignin becomes incorporated into the soil. The actual process of lignin degradation in the soil may be quite different to that occurring in a tree trunk. There is very little direct evidence that any Basidiomycetes degrade lignin in the soil. This may be because we are ignorant of the facts. In studies on soil fungi using standard techniques Basidiomycetes are only rarely recorded. They tend to be slower growing and so are easily overgrown on most culture media. They are often very sensitive to antagonism by others and so suppressed. Most do not produce spores or possess any other readily recognizable cultural features. Nevertheless, they may be equally important as lignin decomposers in the soil itself as they are in litter or decaying wood. The number of soil-inhabiting micro-organisms which have been shown to utilize lignin in the soil is very small, these include a number of bacteria and a few fungi, mainly Deuteromycetes in such genera as *Humicola* and *Phialophora*. The evidence for their ability to utilize lignin has been obtained from the use of chemically extracted lignin and the so-called lignin model compounds which consist of two phenol propanoid monomers linked together by one of the common linkage groups found in the polymer. These have been used incorporated into kaolin pellets to enrich soil, and fungi subsequently isolated from them and tested for their ability to utilize such compounds as vanillic and syringic acids as sole carbon sources. The ability to grow on an utilize these should not be taken as an ability to utilize lignin itself just as the ability to utilize carboxymethylcellulose is not taken as an ability to utilize native cellulose. They are partial degradation products and these fungi should be regarded as occupying a similar niche with regard to lignin as secondary sugar fungi do to cellulose.

Lignin in the soil certainly decomposes very slowly and its degradation there is more of a joint effort between a variety of fungi, bacteria and Actinomycetes. Some of these are capable of cleaving the major bonds between the monomers, others of demethylation and still others of side chain oxidation and so on until the final product enters the respiratory pathways.

As any substratum is decomposed in the soil there is a gradual accumulation of amorphous humus. By far the largest part of this, 50–80%, appears to be made up of highly complex phenolic polymers with simple compounds such as amino acids attached. This is the humic acid fraction which is a very heterogenous one extractable with weak alkali and precipitated with acid. In woodland soils, much of the humic acid may originate from relatively unchanged to considerably modified lignin residues derived in the main from the end products of brown rots. These residues have been degraded to varying degrees by extracellular microbial enzymes and transformed into smaller units which, through further enzymatic or autoxidative reactions, have been converted to humic polymers. Work with tracers has shown that as much as one third of the humic acid in a soil may be derived from lignin and only about one twentieth from cellulose. A further fraction of the humus is derived from other seed plant phenolic based compounds such as the flavonoids. These are phenolics with two aromatic rings and include pigments such as the anthocyanins. These are also degraded to simple phenols which become polymerized into the humic acid fraction. But humus is not solely a product of the degradation of the more resistant parts of seed plants. It is also in part a product of microbial synthesis. When glucose is added to a soil 40–80% of the carbon is lost as carbon dioxide within a few days but even after two years about 5–10% is still present in the soil humus. Intracellular transformation of carbohydrates and other simple organic substances occurs to produce phenols, quinones and other aromatic substances. These are oxidatively polymerized and combined with peptides and other cell constituents to form humic-like pigments, melanins, inside or outside the cell. These serve several functions. They may be deposited in the walls of hyphae, spores or ascocarps to protect against excessive ultra-violet light or as a water-proofing to prevent water loss. Eventually on the death of these structures they become incorporated into the humus fraction.

Humus is extremely resistant to microbial degradation, but nevertheless there is a very slow turnover, the rate depending upon soil type. Humus from a coniferous forest soil in Sweden was C_{14} dated as 370 ± 100 years. A number of fungi have been shown to decompose humic acid in laboratory tests. Humic acid was extracted from a Canadian soil. It contained about one quarter of the total soil carbon and was dated at 785 ± 50 years. It was supplied as the sole carbon and nitrogen source as a 0.2% solution to a number of micro-organisms which had been isolated from that particular soil. Four bacteria and two fungi (*Penicillium frequentans* and *Aspergillus versicolor*) could utilize the humic acid as a sole carbon and nitrogen source. A number of Basidiomycetes including *Coriolus versicolor* (Fig. 3–1) and *Hypholoma fasciculare* (Fig. 3–2c), both active white rot fungi, can also utilize humic acid. Too little is known about the decomposition of humus to be more definitive.

4 Coprophilous Fungi and Fungi of Animal Remains

4.1 Adaptations of coprophilous fungi

Animal dung supports a large fungal flora. Many of the fungi present are always associated with dung. They appear to prefer to grow on dung and are thus termed coprophilous. Although taxonomically unrelated they show a number of common adaptations to their habitat. In many the spore-bearing structures are orientated by phototropism. This is often coupled with a violent spore discharge mechanism so that the spores are dispersed towards the light and away from their staling substratum onto the surrounding herbage. The spore projectile often consists of many spores, for example the entire contents of asci or sporangia. The larger the projectile the less limiting to dispersal is the resistance of the air. It is often mucilaginous so that once impacted it adheres, rather than falls to the soil surface. The spore walls are often pigmented and protect the protoplasm from excessive exposure to sunlight. The spores are ingested with the herbage and survive passage through the alimentary canal of the animal. Many, but not all, may require such a treatment before they will germinate (INGOLD, 1960, 1971).

4.2 The fungal succession on herbivore dung

When fresh herbivore dung is incubated in a damp chamber in the laboratory, a rich crop of Zygomycete sporangiophores, such as those of *Mucor, Pilaira* and *Pilobolus* (Fig. 4–1), develop after 1–3 days and may persist for up to 2 weeks but usually decline after about 7 days. Apothecia of Ascomycetes, such as *Ascobolus, Coprobia* and *Rhyparobius*, appear after

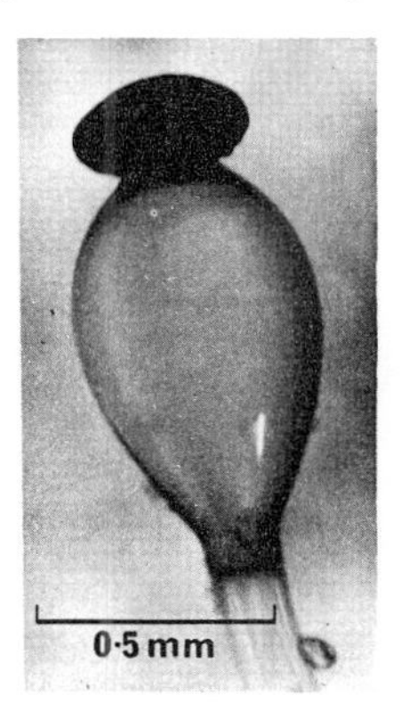

Fig. 4–1 Sporangium and swollen subsporangial vesicle of *Pilobolus*.

5–6 days and persist for up to 3–4 weeks. They are joined after 9–10 days by perithecia of other Ascomycetes, such as *Sordaria, Podospora* and *Chaetomium*. These may last for several weeks and on further incubation basidiocarps of *Coprinus, Stropharia* and *Panaeolus* may appear (Fig. 4–2).

Almost any herbivore dung, on incubation, will produce such a succession with the Zygomycete Mucorales dominating the first phase, followed by Ascomycetes, and leading to a final phase of Basidiomycetes. In the field the same fungal flora develops although weather may interfere with the succession and the sequence may be less regular. For sampling, rabbit or sheep pellets are ideal. They can be incubated in deep plastic or crystallizing dishes on layers of damp filter paper or other absorbent materials. Samples should not be kept in airtight containers, as under such conditions insects and nematodes rapidly break up the dung and anaerobic conditions soon develop. Troublesome insects can be controlled by a proprietary 'fly-kill' aerosol. The fungi are best detected by scanning with a dissecting binocular microscope, although a good hand lens is adequate except for the smaller forms. RICHARDSON and WATLING (1968, 1969) give very useful keys to fungi on dung.

The exact successional pattern, timing and species list varies with the dung. The succession is frequently interpreted as a strict nutritional sequence. In the decomposition of organic matter, such as dung, sugars, starches and proteins disappear first, followed by hemicelluloses, then cellulose, and finally lignin. The Mucorales, which have the capacity of rapid spore germination and hyphal growth, exploit the dung quickly, utilizing the relatively ephemeral soluble carbon sources. They are one of the best examples of primary sugar fungi. When these sources are exhausted, the Ascomycetes utilize the cellulose. These in turn, are replaced by Basidiomycetes, many of which are both cellulolytic and ligninolytic.

The observed succession is, however, one based on the appearance of reproductive structures, either sporangiophores, ascocarps or basidiocarps. The sequence of mycelial development may not be the same. The spores of some, but not all, coprophilous fungi fail to germinate when inoculated without any pretreatment onto fresh dung or even nutrient agar. They develop on the dung from spores which have survived passage through the gut of the herbivore and this process breaks their dormancy. This can be demonstrated by pretreatment of spores with pancreatin for 5 hours at 37°C to simulate this passage. After such treatment the spores of different coprophilous species germinate rapidly, most within 6 hours. After germination, the actual rate of mycelial growth is important in exploitation of the dung. It is found that there is no clear correlation between the growth rates of particular fungi and the observed sequence. It is not a sequence because the spores of the Mucorales germinate more rapidly and their germ tubes and hyphae grow faster than the Ascomycetes, which in turn grow faster than the Basidiomycetes (Table 3).

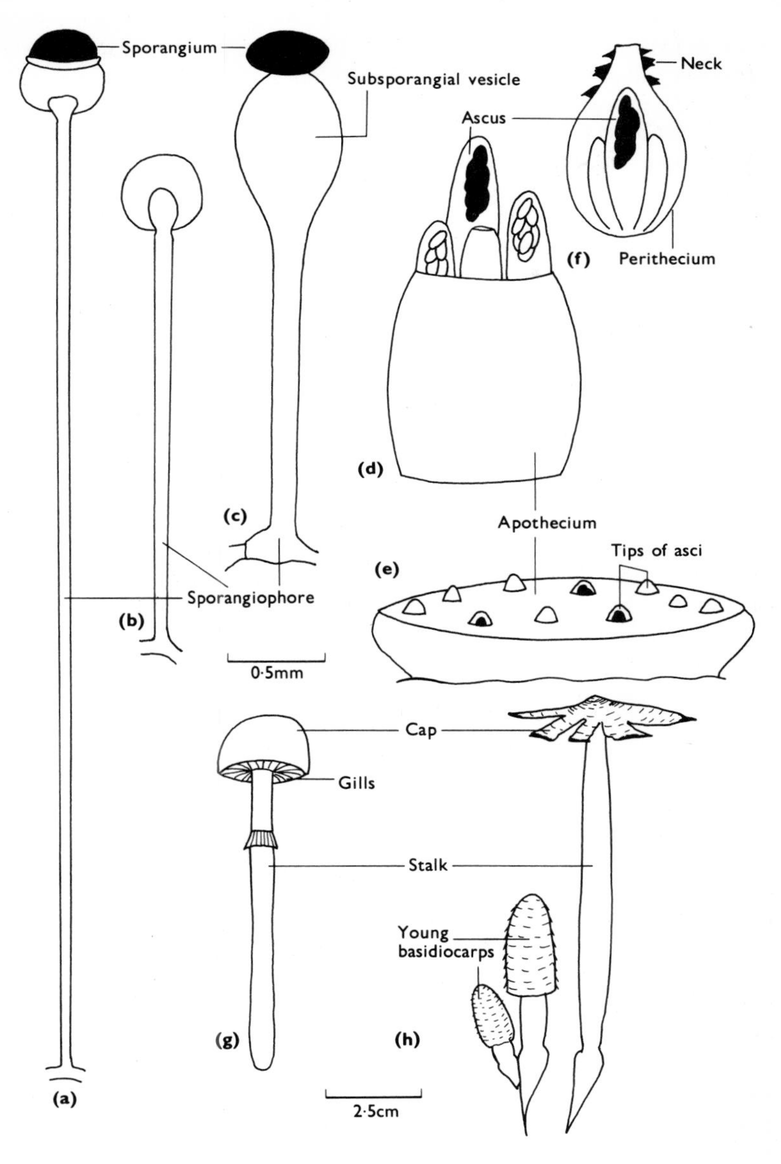

Fig. 4–2 Coprophilous fungi. (**a**) *Pilaira,* (**b**) *Mucor,* (**c**) *Pilobolus,* (**d**) *and* (**e**) *Ascobolus,* (**f**) *Podospora,* (**g**) *Stropharia* and (**h**) *Coprinus.*

Each particular fungus does however take a characteristic minimum time to produce its reproductive structures even if grown on a variety of different substrata, such as sterile dung or standard culture media. To take three examples, *Mucor hiemalis* takes 2–3 days, *Sordaria fimicola* 9–10 days and *Coprinus heptemerus* 7–13 days in culture. This corresponds with the timing on rabbit dung incubated in the laboratory. Thus the minimum time for reproduction, in itself, provides the simplest explanation of the succession. Obviously it is to be expected that it would take an active mycelium of *Coprinus* (Fig. 4–3) longer to amass sufficient reserves to produce its relatively massive basidiocarps than it would take a similar mycelium of *Mucor* to produce its relatively simple sporangiophores.

Fig. 4–3 Basidiocarps of *Coprinus cinereus* growing on a straw compost (×⅔).

In this succession it would appear that the majority of fungi are present when the dung is deposited. They all grow together and initially make preferential use of, and compete for, soluble carbon sources while they last. On their disappearance the Ascomycetes and Basidiomycetes produce hemicellulases and cellulases. By this time, the Zygomycetes, at least, would have reproduced. The Ascomycetes would then reproduce and persist as long as the cellulose lasted. Ultimately all that would be left would be lignin which only the Basidiomycetes could degrade but they begin to produce basidiocarps well before all the cellulose is utilized.

Table 3 Days elapsing before reproduction of coprophilous fungi on incubated dung and latent period of germination, germ tube and mycelial growth rates of spores after pretreatment with alkaline pancreatin and incubation at 100% R.H. (From HARPER, J. E. and WEBSTER, J., 1964.)

	Days elapsing before reproduction	*Latent period* (h)	*Germ tube growth rate* (μm h^{-1})	*Linear growth rate* (mm day^{-1})
ZYGOMYCETES				
Mucor mucedo	2	5–8	18.7	9.1
Pilobolus crystallinus	4	6–9	10.0	4.8
ASCOMYCETES				
Ascobolus glaber	11–12	4–6	44.1	12.0
Rhyparobius dubius	6	6–8	10.8	1.8
Sordaria fimicola	9	4–6	63.7	19.0
Podospora minuta	9–11	4–6	27.5	1.8
BASIDIOMYCETES				
Coprinus heptemerus	9–13	5–8	15.8	3.2
Coprinus patouillardii	37	10–12	9.7	3.7

Competition between microorganisms present in the dung apparently has little effect on the actual time of appearance of reproductive structures but it is important in limiting the duration and intensity of reproduction. Competition for nutrients, the production of antibiotics or some other form of antagonism by other members of the dung micro-flora may limit, for example, the duration of reproduction of the Mucorales. Many of the coprophilous Basidiomycetes, especially species of *Coprinus*, are antagonistic to fungi, such as *Pilaira* and *Ascobolus*. The antagonism can only be observed after contact has actually been made between their hyphae and has been termed hyphal interference. Within minutes of making contact with a hypha of *Coprinus*, cells of *Ascobolus* undergo vacuolation and suffer a loss of turgor. The contact damages the permeability properties of the *Ascobolus* cells but the mechanism is obscure. This can be seen as a very effective form of competition and is probably widespread in fungi in general. It would help to explain the dominance of Basidiomycetes, especially *Coprinus* sp., in the later stages of this succession.

Herbivore dung is another highly complex substratum. It not only contains the comminuted remains of the ingested vegetation but many waste products of the animal in addition to the remains of a very large microbial population, especially bacteria, including cellulolytic ones, from the rumen. These all contribute to the nitrogen content which may be as high as 4%, a three to four-fold increase over the ingested material. There must also be a differential depletion of these nitrogenous compounds and they must have some influence on the growth and reproduction of coprophilous fungi.

With regard to the actual sequence of fungi present, although similar to that on the faecal pellets of insects, the succession on herbivore dung is atypical. Other decomposing plant remains such as deciduous tree leaves, pine needles and grasses amongst others, lack this initial Zygomycete phase (see section 3.6). The first fungi to appear on these are Ascomycetes and Fungi Imperfecti and they are not all true primary sugar fungi in that many possess the ability to produce cellulases. The absence of Mucorales from these and other such substrata may be due, at least in part, to their lower nitrogen content.

A multitude of other factors may also interact to determine this sequence. Herbivore dung also has a high pH. It is usually above 6.5, and, as for pyrophilous fungi, these high values are reflected in the pH optima for growth of coprophilous fungi. These are usually near neutrality rather than on the acid side. This must also have some selective effect on the fungi. Synergistic effects may also be important. *Ascobolus furfuraceus* reproduces better, in culture at least, in the presence of bacteria and *Pilobolus kleinii* reproduces better in the presence of *Mucor plumbeus*. This is due to the release of ammonia by the *Mucor*. *Pilobolus* also requires a growth factor, coprogen, present in dung for growth and reproduction. The coprogen is probably produced by bacteria. Fatty acids are present in herbivore dung and *Pilobolus* makes better growth on these as a sole carbon source than it does on simple pentoses and hexoses. It is the combination and interaction of all these factors which determine the succession.

The coprophilous fungi must possess a wide variety of characteristics if they are to survive and reproduce in such a rich habitat. Like all plant surfaces, the food of herbivores is loaded with a multitude of other fungal spores from the atmosphere, in addition to those of coprophilous fungi. These casual inhabitants either fail to survive passage through the alimentary canal of the herbivore or, if they survive this, they fail in competition with the better adapted coprophilous fungi.

4·3 Keratinophilous fungi

Faecal pellets of predatory birds, such as the hawk, *Falco tinnunculus*, form a rather different type of substratum. The most resistant parts of the pellets are keratin and chitin, rather than lignin. The Mucorales are again the first fungi to appear on the pellets and they are followed by a number of Ascomycetes and Fungi Imperfecti. When most of the readily available nutrients have been removed leaving the more durable parts, such as beetles elytra and animal fur, *Onygena corvina* appears. *Onygena* is a small Ascomycete genus and is peculiarly restricted to animal remains, such as shed sheep and cow horns, rotting hooves and hair. It, like some of the Ascomycetes and Fungi Imperfecti which precede it, is keratinophilous (keratinophilic).

Such keratinophilous fungi are all potentially pathogenic to man, and include the dermatophytes causing diseases of the skin such as 'ringworm'. These are limited to the keratinized parts of the body. They occur widely on the fur of small animals, quills of hedgehogs and feathers of birds. High concentrations can be found in birds' nests. As soil inhabitants they occur on such debris in the soil and they are selectively increased by the addition of fur, feathers and hair.

Hair and degreased animal wool can be used as 'bait' for keratinophilous fungi in suitable habitats, such as the soil. The succession of fungi on hair in the soil is not unlike that on hawk pellets. It is first of all colonized by non-keratinolytic fungi such as *Mucor, Penicillium* and *Trichoderma* sp. Presumably these utilize the more easily assimilable derivatives of the hair. Aqueous extracts of hair include urea, ammonia, amino acids, pentoses and glycogen. These fungi have a high competitive ability in terms of being able to rapidly utilize these less complex nutrients in the substratum. They overlap and are followed by *Chaetomium, Humicola* and *Penicillium* sp. The two former are well known cellulolytic genera and apparently also have the ability to utilize the more resistant parts of the hair. The fungi which follow these are those with the lowest competitive ability but which possess the advantage of being able to degrade the most resistant part of the substratum, keratin, for which there is little competition. These include *Arthroderma* and *Ctenomyces* sp. (Fig. 4–4). These appear to be ecologically restricted to keratin as a substratum because of their low competitive ability in colonizing other substrata. Nevertheless they can utilize simple carbon sources especially where there is little or no competition for these, such as in pure culture.

If hair or wool is to be used as a bait it should be surface sterilized with 1,2-epoxy-propane and cut into 5.00 mm lengths. Soil is a good source for these fungi and a small quantity of the hair should be lightly mixed with a soil sample in a Petri dish and a thin layer of hair dusted over the top. Sterile water should be added at intervals over the 3–4 weeks of incubation to maintain a saturated atmosphere within the dish. Pieces of hair from the surface layer can be removed at intervals and plated out onto potato dextrose or some other nutrient agar. The advantage of using surface hairs is that they are lying on other hairs and fungi growing on these would have actively grown onto the hair from the soil. This reduces the chances of the hair being contaminated with spores lying dormant in the soil, but which nevertheless would grow on the agar. If hairs from within the soil are sampled, they should be washed in repeated changes of sterile water to remove adherent soil particles and its non-active fungal population. Many fungi also sporulate on the surface of the hairs so that direct observations can be used to check on the isolations. Similar fungi will also sporulate on bird's feathers kept in moist chambers. It should

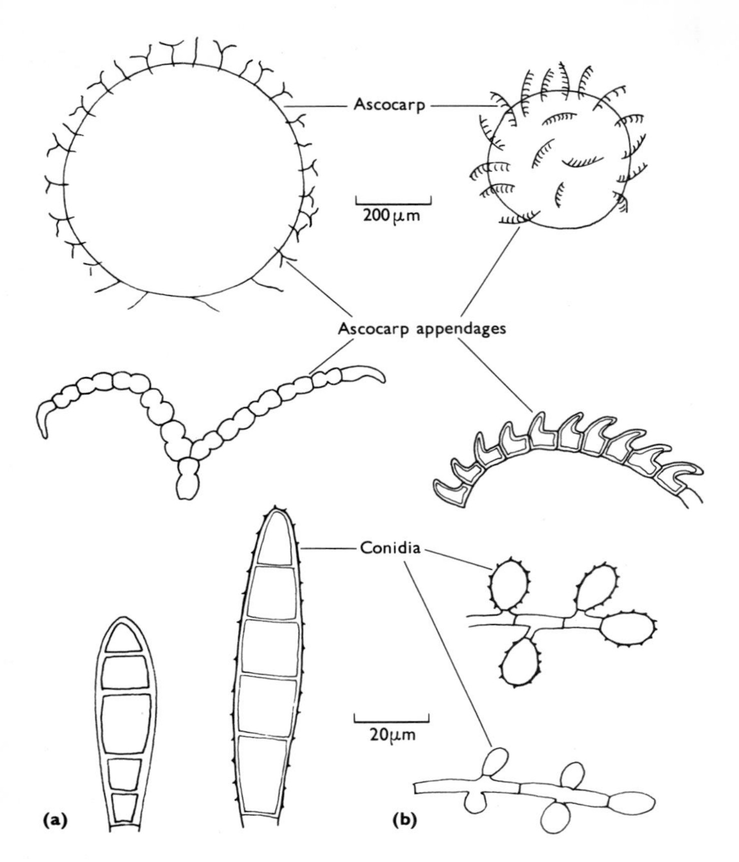

Fig. 4–4 Keratinophilous fungi. Ascocarps, ascocarp appendages and conidia of (**a**) *Arthroderma* and (**b**) *Ctenomyces*.

again be borne in mind that not all the fungi isolated from these keratin-containing substrata are necessarily keratinolytic, because there are sufficient other carbon sources present to meet the requirements for at least limited growth of other fungi.

5 Osmophilous (osmophilic) Fungi and Fungi in Dry Environments

5.1 Moisture requirements

Fungal mycelia, composed of much branched, uniseriate filaments, present a relatively large surface area per unit volume to the influence of the external environment. Such a system has a distinct advantage in permitting rapid absorption of nutrients and water under favourable conditions, but is very susceptible to adverse environmental conditions. Hyphae are very prone to desiccation because they lack a cuticle. Hence perhaps they are usually immersed in their substrata and it is only their reproductive structures which appear on or above the surface exposed to the atmosphere. They depend on there being a certain amount of moisture available for growth within their substrata. Hyphae can also absorb water, and absorb it in sufficient quantity to permit growth, from the vapour phase, if the relative humidity of the atmosphere is sufficiently high.

If fungal growth on any particular substratum is to be prevented a most important environmental factor which must be controlled is the amount of moisture available within it for growth. If this can be kept below a critical level fungi cannot grow. The critical level varies for different substrata because the absolute amount of moisture within each does not usually correspond with the amount available for fungal growth. This latter amount depends upon the physical state and the chemical constitution of the particular substratum. It thus has both tensiometric and osmotic components. Preserves, for instance, have exceedingly high moisture contents but do not go mouldy because of their high concentration of sucrose, which makes the water unavailable to fungi, whereas cellulosic materials such as cotton or textiles will support fungal growth if the moisture content exceeds only 8% on a wet mass basis.

The amount of a moisture in any substratum is a function of the relative humidity of the atmosphere with which it is in equilibrium. Different substrata exposed to a range of atmospheric humidities will take up different quantities of moisture at one and the same relative humidity, but will begin to support growth of a particular fungus at the same relative humidity. The minimum relative humidity permitting growth is definite for any one fungus but varies considerably from one fungus to another. Aquatic fungi are only able to grow in the presence of free water. The Mucorales require conditions approaching saturation and only grow above 93% R.H. Many Penicillia are less restricted and will grow above 80% R.H. and some Aspergilli, members of the *Aspergillus glaucus* and *A. restrictus* groups of species, will grow in as low as 71% R.H. These Penicillia and Aspergilli, together with fungi such as *Cladosporium herbarum* (above 88% R.H.), *Alternaria alternata* (above 89% R.H.) and *Botrytis cinerea* (above 93% R.H.), are what are generally called 'moulds'. We can therefore generalize and say that for most common moulds the minimum relative humidity permitting growth lies between 71 and 95%. Thus a safe limit to impose on most substrata is the moisture content attained when it is in equilibrium with a relative humidity of 70%.

5.2 Osmophilous Aspergilli

Members of the *Aspergillus glaucus* and *A. restrictus* groups are the most xerophytic of all fungi. They are able to germinate and grow at moisture levels where all other fungi are precluded. They occur on all types of organic materials undergoing slow decay at moisture levels just above those at which no decomposition can occur. They have been much studied in the incipient spoilage of stored products, such as grain, foodstuffs, including jams, jellies, salted meat and fish, and leather. Their classic habitat is improperly dried herbarium specimens. They are the green moulds on the inner soles of shoes left in a damp cupboard and the moulds that grow over the surface of jam in an improperly sealed jar.

Tolerance to low humidities is usually associated with tolerance to high osmotic pressures. However they are osmophilous (osmophilic) rather than osmo-tolerant because they make better growth on high sugar containing substrata than they do on low ones. They make no or very slow growth on ordinary culture media, such as 2% malt extract agar, and are thus easily missed in most routine isolation work, but grow very rapidly on the same media supplemented with 40% sucrose. Members of the *A. glaucus* group grow particularly fast on such culture media and are easily recognizable by their bluish-green conidial masses and their minute, but conspicuous, yellow cleistothecia (Fig. 5–1). These ascocarps indicate that they are Ascomycetes in the order Plectascales and in the genus *Eurotium*. The name *Aspergillus*, given to the imperfect or conidial state, is retained by most for convenience and because many Aspergilli in other groups, including the *A. restrictus* group, are not known to produce a perfect state. They can be easily isolated from a number of sources if the right culture media are used. They occur in low frequencies in the atmosphere and can be isolated by exposing Petri dishes of 2% malt extract agar supplemented with 40% sucrose to the atmosphere for 10–20 min on a dry day. On incubation at laboratory temperatures they will be recognizable after 5–6 days and will produce ascocarps in 12–14 days. *Cladosporium* sp. and other moulds also grow on this agar but members of the *A. glaucus* group will overgrow these. A better source are cereal grains treated as suggested below.

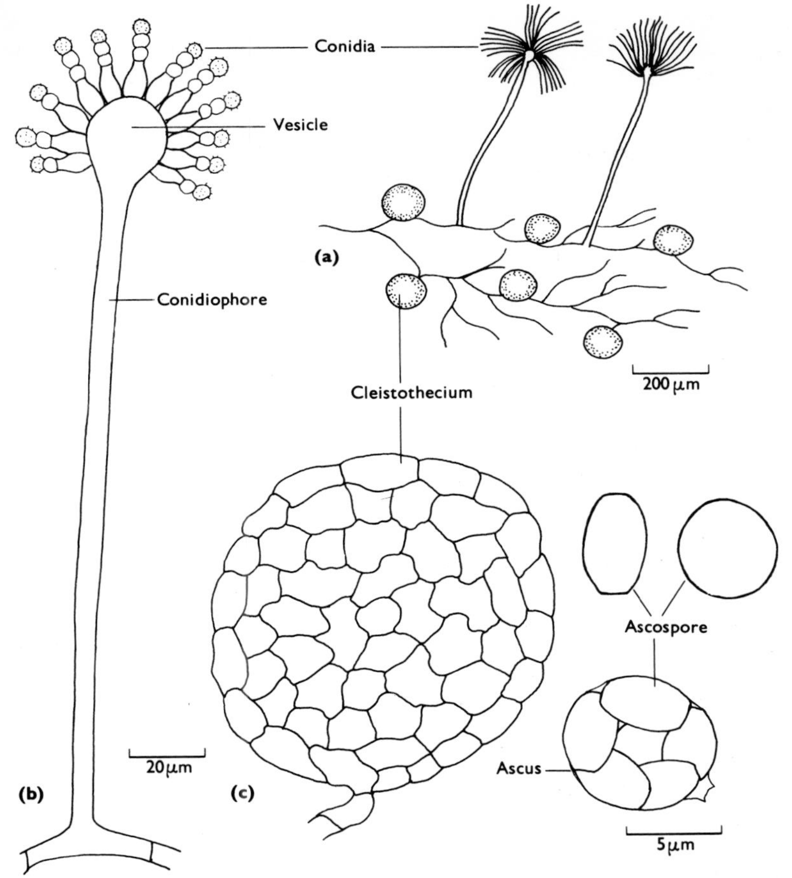

Fig. 5–1 *Aspergillus repens* (*A. glaucus* group). (**a**) habit sketch, (**b**) conidia and conidiophore and (**c**) cleistothecium, ascus and ascospores.

5·3 'Field fungi', 'storage fungi' and stored grain

These osmophilous Aspergilli have been most intensely studied in relation to the spoilage of stored grain and have been designated 'storage fungi' (CHRISTENSEN, 1965). Cereal grains, and many seeds, as they develop in the field and particularly after they mature, but before being harvested, become colonized by 'field fungi'. These include *Cladosporium*, *Aureobasidium*, *Epicoccum* and *Alternaria* sp., the same fungi which colonize deciduous tree leaves and other types of herbage, and also other fungi, such as *Helminthosporium* sp. which are seed-borne pathogens. Most of these fungi require the moisture content of the grain to be in equilibrium with a relative humidity of over 90% before they will grow. This is equivalent to a moisture content of 22–25% on a wet mass basis. The longer the grain is exposed to the weather, the more profusely they develop. If harvest is delayed their presence is indicated by sooty flecks, in particular on the glumes. If it is a good dry harvest, there may be little development of these fungi but they will still be present, with many others, as spores which have been impacted on their surface from the atmosphere. A distinction can be made between fungi present as spore contaminants on the surface and fungi whose spores have germinated and penetrated into the pericarp, by comparing the fungi which grow on culture media inoculated with surface sterilized grain with those inoculated with normal grain.

Grain and most seeds can be surface sterilized by immersion in 70% alcohol for 20 s to wet the surface and removal to a solution of sodium hypochlorite containing 1% available chlorine for 10 min. The grains can then be rinsed with sterile water and placed aseptically onto the surface of nutrient agar, such as 2% malt extract or potato dextrose, just as it is solidifying. It is advisable to add streptomycin (50 μg cm^{-3}) to suppress bacteria to the dishes containing non-surface sterilized seeds. Incubation in the light is also desirable as many of the fungi are dark-spored Fungi Imperfecti. Sporulation of these is stimulated by light. A 'light factor' is important in inducing sporulation in many fungal cultures. The most effective wavelengths are confined to the near ultra-violet region of the spectrum. The light bench as described in the *Commonwealth Mycological Institute's Plant Pathologist's Pocketbook* (1968) is a very useful asset. This utilizes near ultra-violet 'black light' and daylight tube lighting. If another sample of non-surface sterilized grains are plated out onto 2% malt extract plus 40% sucrose agar, they will develop colonies of members of the *Aspergillus glaucus* group.

After harvest cereal grains are usually commercially stored at below 13.0% moisture content, equivalent to a relative humidity below 70%. At this level the field fungi can no longer grow. Spores of the storage fungi will be present on the grain as casual inhabitants. They do not invade the grain to any extent before harvest. Nor do they grow at such low moisture contents, but should small moisture pockets occur in the stored grain giving local moisture contents of 13.2% or just above, these storage fungi begin to grow and colonize. Each species has its own lower limit at which it will grow. The most xerophytic of all are *A. halophilicus* and *A. restrictus*. They begin to grow between 13.2 and 13.5%. They do not grow rapidly but produce moisture by their respiration and slowly increase the overall level. When the level reaches 14.0–14.2% members of the *A. glaucus* group begin to grow and likewise increase the overall moisture content enabling still less xerophytic fungi to colonize. Their metabolic activities also bring about heating which in addition to increasing the rate of deterioration encourages the development of thermophilous micro-organisms. As a result of all their activities, the percentage germination of the grain falls. It becomes discoloured and biochemical changes occur in the endosperm so that it is not fit for food. Toxins may be produced in sufficient quantities to constitute a health hazard, if eaten. The final result may be complete spoilage.

The presence of these storage fungi in grain spoilage reflects their ability to grow where others cannot, rather than their presence as a heavy spore load on freshly harvested grain. Their real importance lies in their ability to initiate growth at minimum moisture levels, establishing bridgeheads and facilitating colonization by less xerophytic fungi. They have a competitive advantage over the latter whose spores will not germinate until the moisture stress is less severe.

On grain as a substratum they illustrate how a change in an all important environmental factor, in this case moisture content, can bring about replacement of one group of fungi, 'field fungi' by another, 'storage fungi'.

5·4 Fungi as xerophytes

It has been stressed that the osmophilous Aspergilli are the most xerophytic of fungi because they are able to germinate, grow and reproduce at moisture levels where all other fungi are precluded and that the vegetative hyphae of other fungi are very susceptible to desiccation. The latter would survive such desiccating conditions by the production of thick-walled resting spores or multicellular resting structures, such as sclerotia. These latter, like all complex structures in fungi, are formed by hyphal aggregation and fusion. They are often more or less globose and differentiated into an outer thick-walled pigmented rind, enclosing larger thin-walled cells, often filled with a food reserve, such as glycogen or oil. They remain viable for longer periods than hyphae and, like spores, only become active when moisture is available.

Being usually immersed in the substratum, the hyphae may to some extent be protected from extreme environmental conditions. It is only the reproductive structures that project into the drier aerial environment. This section considers how some of these endure drought.

The discharge of ascospores and basidiospores involves the activity of turgid cells, so that in Ascomycetes and Basidiomycetes in particular, spore discharge necessitates a supply of water to the layers of the reproductive structure where they are formed. Under dry atmospheric conditions many ascocarps and basidiocarps lose water rapidly and shrivel and spore discharge ceases. Most basidiocarps of the toadstool-type such as those of *Amanita*, *Boletus* and *Coprinus* are somewhat fleshy and are unable to withstand drought. They lose water very easily to the atmosphere, shrivel and do not recover on return to more humid conditions. A notable exception to this are the basidiocarps of *Marasmius oreades*, the fairy ring fungus. These can be air dried and will shrivel up, but will function normally again, by discharging basidiospores, after rewetting. In this respect it is like the majority of the tougher, more leathery or corky-textured, bracket-shaped basidiocarps of the polypores and the minute flasked-shaped ascocarps of many Ascomycetes. These may be considered as drought-enduring xerophytes. They discharge spores only in conditions of high humidity and in dry conditions shrivel and become inactive but rapidly recover on return to high humidities.

The best example of this ability to dry out is seen in *Schizophyllum commune*. This grows on dead wood, such as branches of beech, and produces small, 1–3 cm, laterally attached, whitish fan-shaped basidiocarps. The upper surface is covered by down-like hyphae and on the lower gills, bearing the basidia, radiate from the point of attachment. They are of different lengths and are divided into two vertically. In moist conditions the gills hang vertically but in dry conditions the gill halves curl outwards so that the basidia become enclosed in a chamber and are no longer exposed (Fig. 5–2). This may help to check the rate of water loss

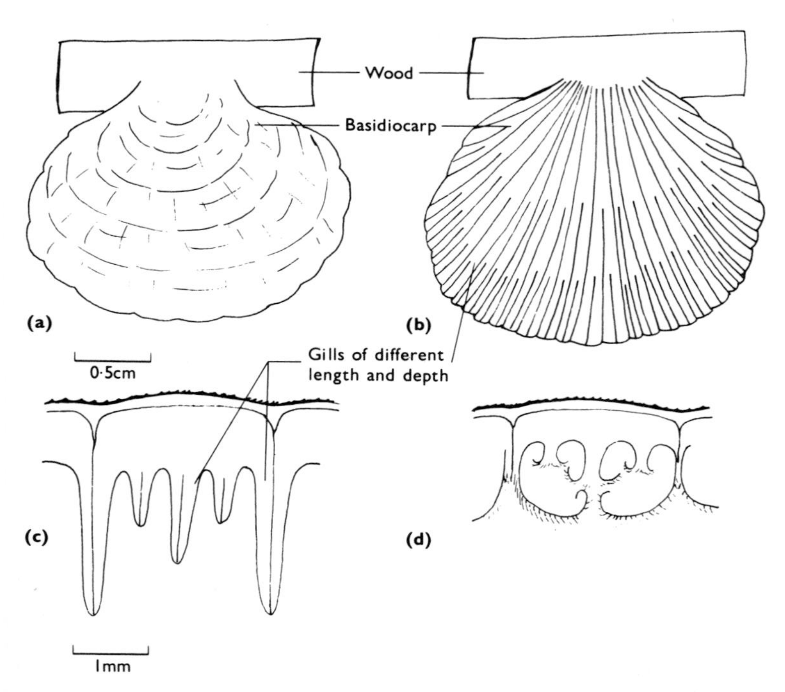

Fig. 5–2 *Schizophyllum commune*. (**a**) basidiocarp from above and (**b**) from below, (**c**) vertical section of gills in a moist atmosphere and (**d**) in a dry atmosphere.

during the initial phase of drying. The whole basidiocarp can, however, be air dried and kept for 2–3 years. On re-wetting the gills uncurl and viable basidiospores are soon discharged. It falls into that small category of plants, such as *Selaginella lepidophylla* and *Ceterach officinarum*, the Rusty-back Fern, which are commonly called 'resurrection plants'. They can dry out to a remarkable degree but revive on re-wetting.

By contrast the compound ascocarps of *Daldinia concentrica* (Fig. 5–3 and 3–2a) can continue to discharge their ascospores for some time, 3–4 weeks, under dry conditions. They do this by possessing a water reserve. The fungus produces its ascocarps in large hemispherical stromata on branches of ash. They are some 2–5 cm in diameter and the individual ascocarps, perithecia, are formed just beneath the outer reddish black rind. The remainder of the stroma consists of concentric zones of sterile tissue, which is essentially a water reserve. On drying this is slowly used, there is no decrease in overall volume but the density falls. Such a system enables *Daldinia* to discharge its spores throughout the drier summer months. In some other Ascomycetes the ascocarps contain a rich supply of jelly-like substances. These again act as water reserves and are used to sustain spore discharge under dry conditions (INGOLD, 1971).

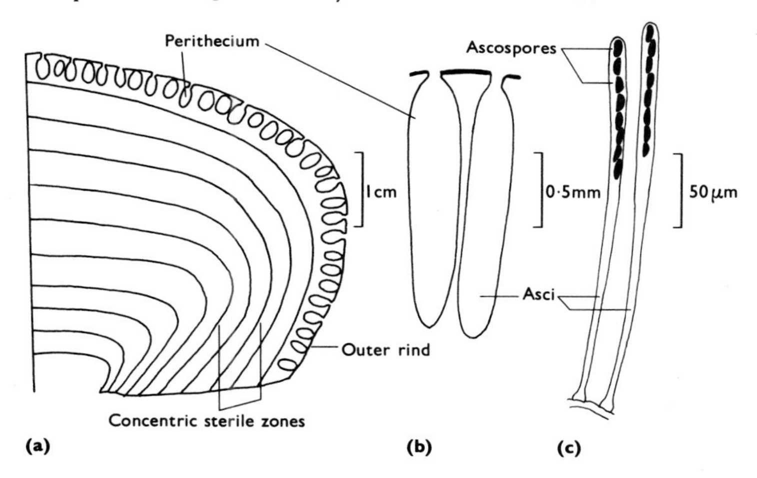

Fig. 5–3 *Daldinia concentrica*. (**a**) vertical section of half of a perithecial stroma, (**b**) vertical section of two perithecia and (**c**) asci.

There are a few Basidiomycetes in which the basidiocarps continue to liberate spores even in more prolonged periods of drought. These all produce large, often perennial bracket-shaped basidiocarps, such as those of *Ganoderma applanatum*. They all grow on wood and basidiocarps isolated from wood soon cease spore discharge as they do not have a water reserve. They obtain their water from the wood. This can be both water present in the wood and water released by the mycelium in degrading the wood. In the latter respect they behave like *Serpula lacrymans*.

6 Aquatic Fungi

6.1 Aquatic Mastigomycetes

Many aquatic Mastigomycetes are widespread and very common saprophytes, or facultative parasites, on all varieties of plant and animal debris including algae, pollen, seeds, fruits, twigs, dead insects, exuviae of insects and so forth. They occur in almost every kind of fresh water habitat. Some are marine and many more occur in soils. They never develop in abundance in nature to be collected in quantity as one might collect filamentous algae, such as *Spirogyra*. However, their presence can be revealed simply by providing them with an appropriate substratum in the form of a 'bait', which provides a small, concentrated food supply that is relatively insoluble. This rapidly becomes colonized by free-swimming zoospores. As a group they exhibit a very wide diversity of thallus form. They all characteristically possess zoospores with either one posterior or anterior flagellum or two anterior or lateral flagella. The first part of this chapter considers the habitats of and the methods used to isolate some of the commoner members.

6.2 Chytrids

The simplest aquatic Mastigomycetes are the Chytridiales or Chytrids. The majority of these consist of either a unicellular thallus which is entirely converted into a zoosporangium or resting spore or they possess one or many fine vegetative rhizoids in addition to the single reproductive structure. All possess zoospores with a single posterior flagellum. They differ from most other Mastigomycetes in that many are cellulolytic, while others grow on keratinous and chitinous substrata.

They are not all truly aquatic and although they occur predominantly in fresh water, any soil has a regular and constant flora of Chytrids. Soil from the humification layer of a deciduous wood is a good source. If about 3 g of soil are placed in a Petri dish, barely covered with sterile water and suitable baits added, Chytrids will begin to appear within 2–3 days at 20°C. Good baits for cellulolytic Chytrids include cellophane and strippings from succulent bulb scales of onions. Both of these are translucent and excellent for microscopic examination. Cellophane is prepared by cutting a thin sheet into 10.0 mm^2 and boiling in distilled water to remove any plasticizers before autoclaving in distilled water. Onion strippings are boiled in water to remove most soluble substances and then autoclaved in distilled water. Such baits are also suitable for baiting pond or stream water. From these waters many Chytrids can be collected on exuviae of caddis flies, mayflies, midges and other insects and large populations can be obtained by baiting the water with purified chitin. For keratinophilous Chytrids, hair also affords a good bait.

One of the most widespread of all Mastigomycetes and certainly one of the most conspicuous, because of the rose-colour that it imparts to substrata on which it is abundant, is the Chytrid *Rhizophlyctis rosea* (Fig. 6–1). This is essentially a terrestrial Chytrid in that it is very common

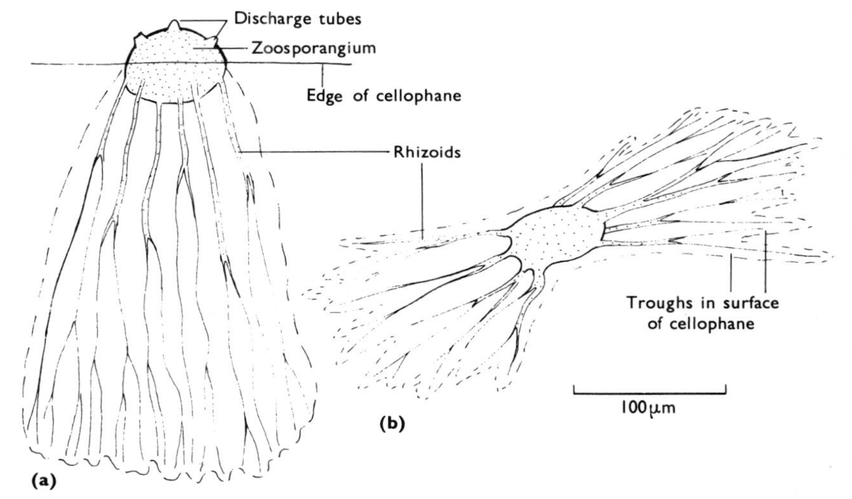

Fig. 6–1 Chytrids. (**a**) mature zoosporangium and rhizoidal system and (**b**) developing thallus of *Rhizophlyctis rosea* on cellophane.

in soils, but is never found in permanently submerged muds. It is most easily isolated by preparing Petri dishes containing mineral agar with an ammonium nitrogen source at about pH 7.0 (Table 4) and covering the surface with a sterile 90.0 mm filter paper. If the centre is inoculated with a few crumbs of soil from a field or hedgerow and the inoculum dispersed by adding just sufficient sterile water, under 1 cm^3, to wet the filter paper, scattered bright rose-coloured colonies will appear after 5–6 days at 20°C. Microscopic examination will reveal the large (up to 250 µm diam) zoosporangia with rose to orange-coloured contents, each with several broad discharge tubes and an extensive rhizoidal system arising from a number of points. It would appear to be an important cellulolytic fungus in many moist soils.

A great variety of other substrata have been used as baits for Chytrids and various investigators have allowed their imagination free rein. One other excellent bait is pollen. As a bait it has many advantages. It can be dusted onto the surface of water over soil where it floats and is easily recoverable by using a wire loop. The Chytrid thalli which develop on pollen are smaller and more typical in size than those of *Rhizophlyctis rosea*.

Table 4 Constituents of mineral agar and sea water–glucose agar.

Mineral agar (after STANIER, R. Y., 1942)	
Potassium phosphate (K_2HPO_4)	1.0 g
Ammonium sulphate ($(NH_4)_2SO_4$)	1.0 g
Magnesium sulphate ($MgSO_4 . 7H_2O$)	0.2 g
Sodium chloride (NaCl)	0.1 g
Calcium chloride ($CaCl_2$)	0.1 g
Ferric chloride ($FeCl_3$)	0.02 g
Agar	15.0 g
Tap water	1.0 l
Sea water–glucose agar	
Glucose	1.0 g
Yeast extract	0.1 g
Agar	15.0 g
Sea water	1.0 l

50 μg cm^{-3} of streptomycin should be added, after autoclaving, to suppress bacteria.

Zoosporangia are produced over the surface and their rhizoidal system develops within the pollen grain. Pine pollen has most often been used mainly because it can be collected in such quantity in May and easily stored dry. If a small quantity of soil from a pine woodland or almost any site is immersed in sterile water in a Petri dish and the water surface dusted with pollen, species of *Rhizophydium*, especially *R. pollinis-pini* and *R. sphaerotheca* (Fig. 6–2), will develop within 3 days. A transfer of a wire loopful of such pollen to fresh pollen over sterile water will result in colony numbers of epidemic proportions in a matter of days.

6.3 Water moulds

The best known and most ubiquitous aquatic Mastigomycetes are the Saprolegniales, the water moulds. Like the Chytrids, many have been isolated from soils but well-aerated ponds are ideal habitats. For these, the best baits are house flies, fruit flies, ants' eggs or any small seeds, such as those of *Brassica* sp. Caryopses ('seeds') of grasses, such as those of *Agrostis* and *Lolium* sp., are good baits. The flies and ants' eggs should be boiled for 2–3 minutes and the seeds until the coats split. All baits should be transferred to large dishes of pond water of sufficient depth to barely cover. The amount of bait should be kept at a low level in relation to the volume of water used. One or two per 50 cm^3 of water is sufficient. If too much bait is used excessive bacterial growths are encouraged. Gentle aeration helps to avoid this.

The Saprolegniales, unlike the Chytridiales, are true mycelial fungi and within 2–3 days, the coarse, characteristically aseptate and rapidly

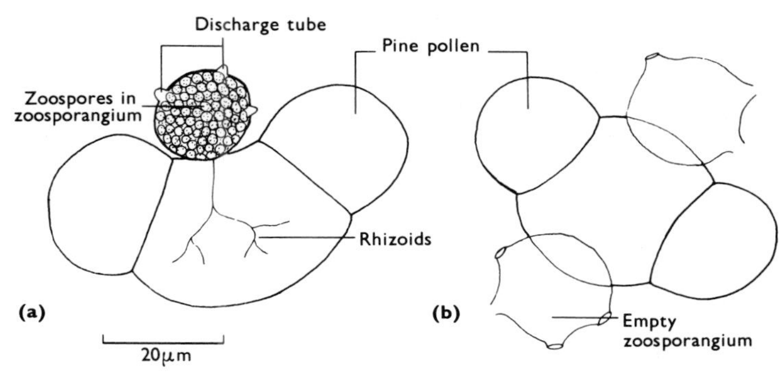

Fig. 6–2 Chytrids. (**a**) mature zoosporangium and (**b**) discharged zoosporangia of *Rhizophydium sphaerotheca* on pine pollen.

growing hyphae appear and develop radially from the bait. Cylindrical zoosporangia develop at the tips of hyphae and produce copious and very active biflagellate zoospores. With age, sexual stages may develop (Figs. 6–3 and 6–4). Although they are difficult to free from bacterial contamination, they can be cultured more readily than most Chytrids. Corn meal agar is a good natural medium and any synthetic media used must be low in free sugars and contain both an osmotically inert and slowly hydrolyzable carbon source, such as starch, and organic nitrogen. They are all non-cellulolytic and occur on submerged plant and animal debris, especially dead fish, during the initial stages of decomposition. They are aquatic primary sugar fungi.

6.4 The Blastocladiales and Monoblepharidales

Representatives of two other orders of aquatic Mastigomycetes can be obtained by baiting procedures but with these it is more profitable to use bait in the field. Members of the Blastocladiales can be obtained by baiting streams with various fruits. Apples, rose hips and tomatoes are good baits. The fruit should be firm and slightly under ripe with an undamaged skin. The fruit should be placed in a small, weighted galvanized metal or wire basket, submerged in about 30 cm of water. The time of collection of the baits will vary with the water temperature but 4–6 weeks submersion is usually sufficient to produce a good crop of Blastocladiales. On removal, the fruit should be washed under a tap to remove surrounding slime and the white, granulated hemispherical pustules of densely compacted thalli will be more clearly revealed. The material should ideally be examined immediately after collection, since, when placed in fresh water, the changing environmental conditions often induce rapid zoospore production and release. Alternatively they can be kept for a limited period submerged in water in a refrigerator. Each pustule is usually composed of a single species, although it it not uncommon to find 2 or 3 species together in each pustule. Members of the two genera *Rhipidium* (Leptomitales) and *Gonapodya* (Monoblepharidales) closely parallel these *Blastocladia* in form and may occur mixed with them. If individual pustules are picked off, teased gently with needles in a drop of water, carefully washed to remove bacteria, and examined microscopically, typical thalli of species of *Blastocladia* will be seen (Fig. 6–5). These have a basal trunk cell which is anchored to the substratum by branched tapering rhizoids which are the only parts penetrating the fuit. The apex of the trunk cell may be swollen and may bear over its surface cylindrical zoosporangia and ovoid thick-walled resting sporangia or it may bear dichotomous branches with both types of sporangia at their tips.

We know very little about the ecology of these fungi other than that they occur widely in fresh water, especially relatively still, shallow streams.

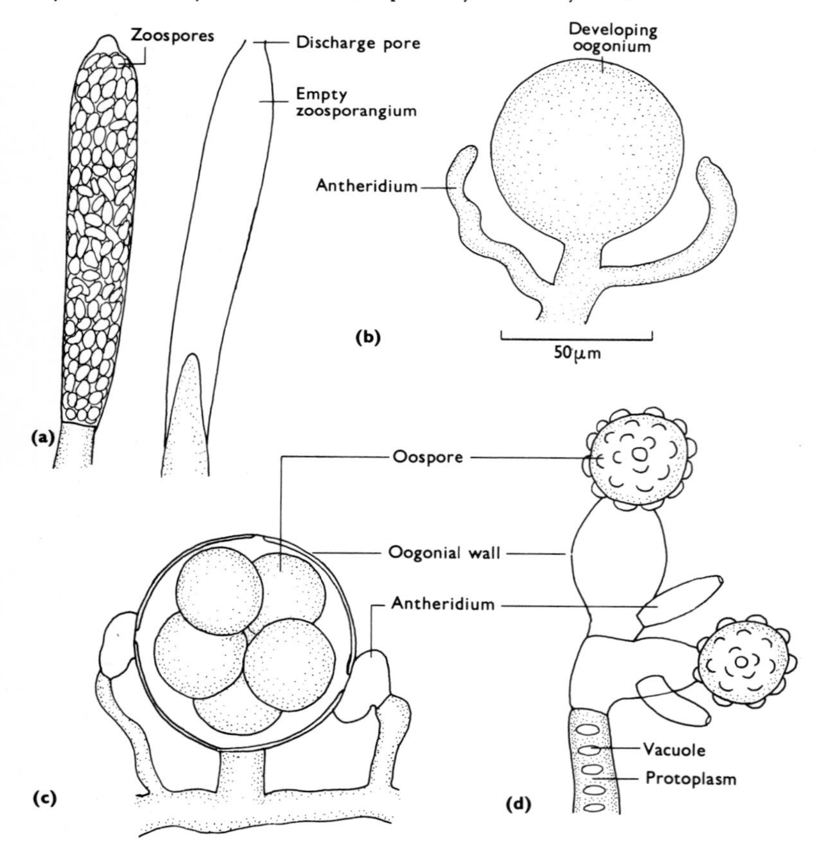

Fig. 6–3 Water moulds. (**a**) a mature and a discharged zoosporangium, (**b**) a developing oogonium and antheridia and (**c**) mature oogonium with oospores of *Saprolegnia*, and (**d**) oospores of *Monoblepharis*.

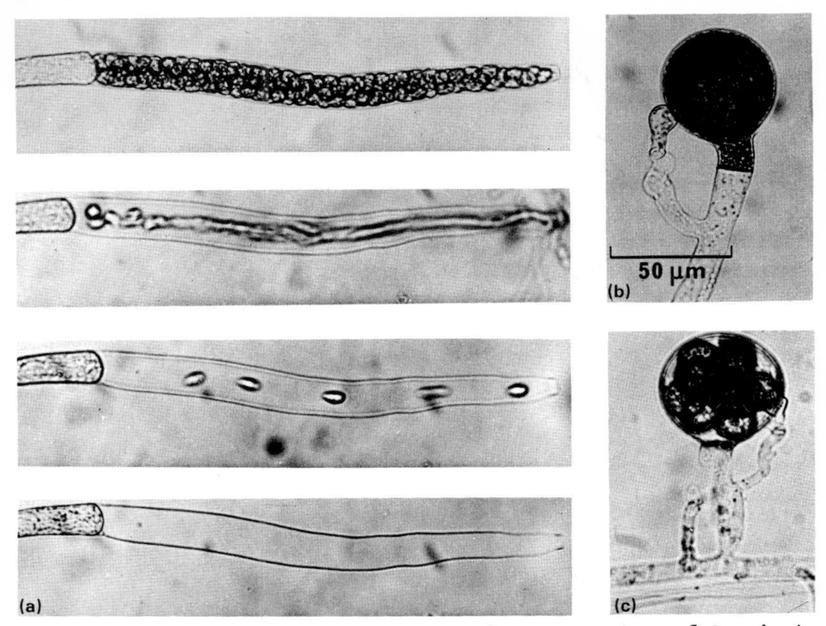

Fig. 6–4 (**a**) Sequences in the discharge of zoosporangium of *Saprolegnia*. (**b**) A developing oogonium and antheridium of *Saprolegnia*. (**c**) A mature oogonium of *Saprolegnia* with oospores.

Waterlogged twigs of deciduous trees are also good sources for these but provide better substrata for the Monoblepharidales. Most of these latter fungi produce delicate hyphae rather than compact thalli. The hyphae are easily recognizable by their characteristic vacuolation. The protoplasmic contents form a network of finely granular strands crossing the hyphae at right angles and enclosing vacuoles of a uniform size. In profuse growth the fungi are visible as pale grey tufts. When sexual reproduction occurs these turn yellowish brown, due to the presence of very many golden oospores (Fig. 6–3). Asexual reproduction occurs by the production of narrow, cylindrical zoosporangia borne terminally at the tips of hyphae. Like those of the Chytridiales and Blastocladiales, these produce posteriorly uniflagellate zoospores.

All members are saprophytes on waterlogged, entirely submerged twigs of deciduous trees which still retain their bark. They occur in shallow ponds with a fresh water supply and in streams under 30 cm in depth and in which the water is almost still and free from silt. The fungi colonize via the lenticels and silt blocks these. On collection, suitable twigs should be placed in a dish, covered with distilled water and placed in a refrigerator at about 3°C. Such material rarely shows visible growths when it is first brought in. The low temperature reduces the growth of many of the larger, less delicate Saprolegniales which at laboratory temperatures tend

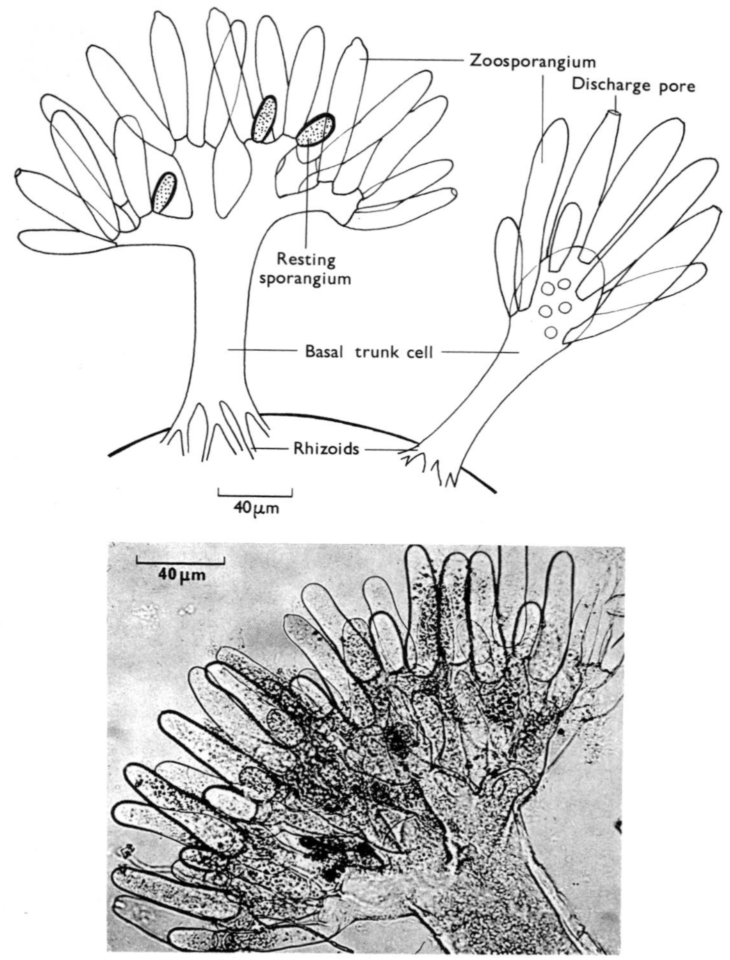

Fig. 6–5 Thalli of *Blastocladia*.

to crowd out the slower growing Monoblepharidales. Twigs collected during the spring and autumn usually produce vigorous growths in 2–4 weeks, whereas those collected in the summer and winter months take 1–4 months. This can be related to the dormant periods of their oospores. The oospores remain dormant over the winter and do not germinate until March, April or May when there is rapid vegetative growth and spread. Oospores are soon produced and the colonies die down. These oospores remain dormant until October and November when there is another flush of growth. Hence, in any one year there are two periods of germination and growth, one in spring and another in autumn with resting periods during the summer and winter months.

6.5 Nutrition and water pollution

There is still a great deal to be learnt about the nutrition of these fungi. There is virtually no data available for the Monoblepharidales because so few have been cultured. Only some of the Chytridiales can utilize carbohydrates as complex as cellulose. They are the only order in which some members can utilize nitrate nitrogen and in which all can reduce oxidized forms of sulphur, such as sulphate, for their sulphur requirements. The others require sulphur in the reduced form. Some Chytridiales can synthesize all their essential vitamins, whereas the Blastocladiales have lost the ability to synthesize the pyrimidine moiety, if not the complete thiamin molecule. These few facts are of little value in attempting to evaluate their respective roles in the decomposition of organic materials in aquatic environments.

Leptomitus lacteus, the 'sewage fungus' has already been mentioned. It is autotrophic for vitamins, requires organic sources of nitrogen and is unable to assimilate sugars and thus must be supplied with other carbon sources, such as amino acids and fatty acids. Whereas pyruvate and lactate are both acceptable carbon sources, alanine or leucine can serve as sole carbon and nitrogen sources. In nature such substances are present in water polluted by sewage and such micro-organisms as *L. lacteus* and the filamentous bacterium, *Sphaerotilus natans*, may be of possible value as biological indicators of pollution. Increasing populations of these in a particular water will indicate increasing levels of sewage pollution.

In all these aquatic Mastigomycetes the zoospore is the principal unit of dispersal. Their whole evolutionary history has taken place in water but there are a large number of aquatic fungi which, so it would appear, have migrated from land to water. These are mainly Ascomycetes and Fungi Imperfecti, but there are also a few Basidiomycetes. Some of these will now be considered.

6.6 Aquatic Hyphomycetes

It was stressed that the fungi which occurred on faecal pellets of millepedes which fed on deciduous tree leaves were different to those which occurred naturally on such leaves in the litter (section 3.10). If leaves fall into water and become submerged, the fungi which subsequently develop are again very different. At leaf-fall, the leaves are already colonized by a variety of leaf saprophytes. If they should fall into small, clean, well-aerated streams flowing rapidly through wooded country, these fungi are rapidly replaced by a characteristic microflora of Fungi Imperfecti, the aquatic Hyphomycetes.

Aquatic Hyphomycetes grow particularly well on all sorts of leaves of deciduous trees and shrubs. They are found on leaves after they have been submerged for several weeks by which time they have turned brown and are becoming reduced to skeletons. They can be found throughout the year but are most common from late September to December, when fallen leaves are most abundant in the stream bed. The most striking characteristic of these fungi is their spores. The conidiophores are single hyphal filaments which project into the water and bear at their apex colourless spores often of an unconventional form. The majority have branched spores with the commonest type being ones with four arms diverging from a common point (Fig. 6–6). These are called tetraradiate conidia and a study of their modes of development shows considerable diversity in the different genera, suggesting that parallel evolution has occurred and that this spore has some survival value in the aquatic environment (INGOLD, 1960). Vegetative growth, spore development, liberation and dispersal takes place under water. In streams in which they

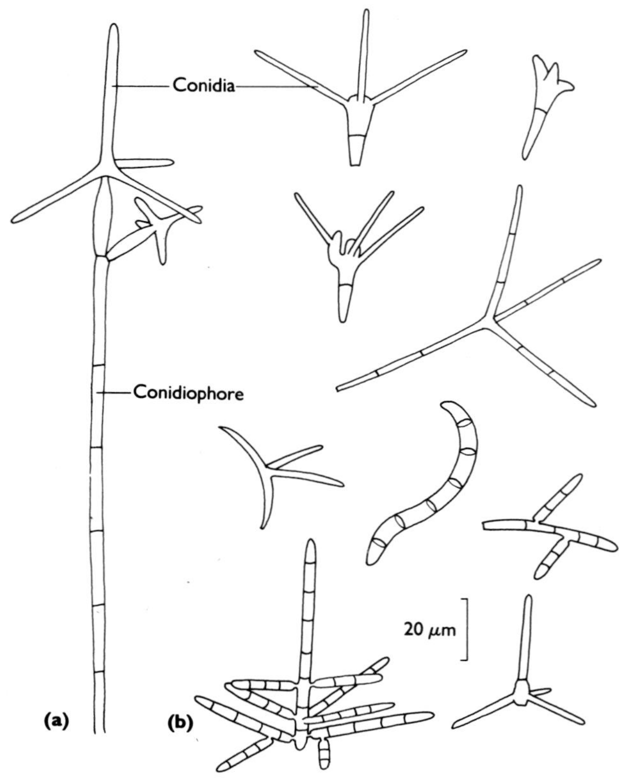

Fig. 6–6 Aquatic Hyphomycetes. (**a**) conidia and conidiophores of *Lemonniera* and (**b**) a variety of conidia from scum samples.

occur foam or scum collects at barriers to water flow, such as fallen branches, or below waterfalls. The foam acts as a very efficient spore trap. If samples are ladled off with a spoon into small screw-topped vials or specimen tubes and examined microscopically, a wide range of spore types will be seen. It is advisable to add an equal quantity of formalin-acetic alcohol to the scum sample to kill and fix the spores, otherwise they may germinate during transit to the laboratory. Not all the spores will be tetraradiate, some will be of a more conventional form, others sigmoid and still others much branched. Spore numbers in rivers and streams can be estimated by filtering 250–500 cm^3 samples of water through Millipore filters (8 μm pore diam.) The filters should be sucked dry, placed in Petri dishes, covered with 0.1% cotton blue in lactophenol to kill and stain the spores and then heated for 45 min. at 50–60°C. The heating makes the filters sufficiently transparent to enable the spores to be recognized using the low power objective of a microscope. The spores, unlike the zoospores of Mastigomycetes, are of sufficient size and have such distinctive features that they can often be identified to the specific level. INGOLD (1975) has produced a very easy to use and comprehensive guide to these fungi. Spore numbers may be high. In the River Exe near Exeter in October the numbers vary from 1000–7500 l^{-1}. Spore development can be followed microscopically on leaf fragments. Partially skeletonized leaves should be collected, washed thoroughly and placed in a developing or dissecting dish filled to about 4 cm in depth with water, through which air is gently bubbled for 24 hours. If fragments are then mounted in water and covered, conidiophores will be observed producing spores in the interveinal spaces of the leaf skeletons.

Over 60 of these aquatic Hyphomycetes have been described. It would appear that the special biological value of such a spore form in an aquatic environment is its ability to act as a minute anchor for attachment to a suitable substratum. On impaction the spores make a stable three-point landing. Each of the three arms rapidly produces an attaching appressorium and germination takes place immediately. Attachment is a real problem for any spore in a stream.

Although a great deal is known about the morphology and development of these fungi, rather less is known of their ecology. They are undoubtedly much more common and abundant, in terms of spore numbers, in rapidly flowing streams than they are in slow flowing ones. The faster flowing ones are more turbulent and it appears that turbulence not only detaches spores more rapidly from their conidiophores but also stimulates the hyphae to branch and produce more conidiophores. Both these features lead to higher spore production. They are not all strictly confined to the aquatic habitat and may be found on leaves well away from water. If fallen deciduous tree leaves from the woodland floor are submerged in distilled water for 4 days at 10°C in a Petri dish they will often produce a few spores. It seems that they may grow elsewhere but sporulate better under turbulent conditions in freshwater.

A few have been demonstrated to have perfect states in the Ascomycetes and at least one in the Basidiomycetes. Some of these are terrestrial forms. This amphibious condition would be the ideal with water favouring the development of the conidia and terrestrial conditions the development of their perfect states with wind-blown ascospores or basidiospores. This would help to explain how they are dispersed from stream to stream and how they manage to progress upstream in spite of water flow.

These fungi are invaluable members of such freshwater systems as they play a key role not only in the decomposition processes but also as intermediaries in the food chain. They form the most substantial fungal element in well-aerated streams and their major substratum, deciduous tree leaves, forms the most substantial component of organic matter annually added to such systems.

Many streams have an extremely low primary productivity. This may be especially so if they run through woodlands and are partially shielded from the direct rays of the sun by the trees. However, they are often well supplied with substantial amounts of organic matter in the form of dead leaves and twigs from the adjacent terrestrial vegetation. It has been estimated that a stream in a wooded valley receives at least 1 kg of leaves per metre of its length each year. Thriving stream communities develop largely dependent upon this added organic material for its food supply. It may contribute anything from 50% to 99% to the total energy budget of stream communities. As in terrestrial communities only a very small fraction of the energy locked up in leaf material can be directly exploited by the fauna. For example, the common freshwater amphipod, *Hyalella azteca*, assimilates only 5% of the material it ingests when fed on elm leaves. Thus, in this respect, it is comparable with the terrestrial caddis fly and the millepede. To enable them to gain access to the remaining energy, the fauna depend upon the activities of micro-organisms. These degrade cellulose and other polymers and convert these to their own biomass. Microbial carbohydrates, fats and proteins are then digested by the fauna. It has long been recognized in terrestrial communities that fungi as well as bacteria are important intermediaries in the food chain but until quite recently it was assumed that in aquatic environments, especially streams, bacteria were the exclusive or at least the predominant intermediaries. This is in spite of the fact that the hyphal organization of fungi gives them an advantage over bacteria in the utilization of bulk cellulose, the major polysaccharide in leaves.

Under laboratory conditions and using antibacterial or antifungal antibiotics, or both, to suppress bacteria or fungi, or both, growing on elm leaves collected from streams, it has been shown that the loss in dry mass of leaves in such systems was highest when the fungi were allowed to grow. The increase in protein content which normally occurs in leaves when they decay in freshwater was also significantly depressed in the presence of antifungal antibiotics but was not significantly changed in the presence of antibacterial ones. From this it can be concluded that these fungi bring about substantial protein increment in decaying leaves and that they are more successful in degrading leaves than are bacteria.

A number of freshwater detritus-feeding fauna have been shown to prefer partly decomposed leaves rather than sterile or freshly fallen ones. They apparently prefer leaf tissues being degraded by fungal hyphae rather than leaf tissues alone. The fungal hyphae may be the attraction. In one study a comparison was made in the efficiency with which the amphipod, *Gammarus pseudolimnaeus*, converted leaf tissues and fungal mycelium into its own biomass. Different groups of animals were given as their sole food supply either maple (*Acer*) or elm (*Ulmus*) leaves or the mycelium of one of ten fungi – five terrestrial ones and five aquatic Hyphomycetes. The actual amount consumed by those on the leaf diets was in the order of ten-fold that of those on the fungus diets but the greatest weight increases were found on those feeding on some of the fungi especially the aquatic Hyphomycetes. The fungi represent a form of nourishment about ten times as concentrated as leaf material. Thus, prolific fungal growth on leaves will very substantially improve their food value to the stream fauna but the fungi also increase the palatability of the leaves. Gammarids when given the choice will actually select leaves colonized by aquatic Hyphomycetes in preference to sterile leaves.

For such detritus feeders, fungi may fulfil two functions. They may be a source of major nutrients or they may supply essential growth substances such as B vitamins and sterols otherwise deficient in the animals' diet. *Gammarus* and other detritus feeders ensure a supply of these substances by preferentially feeding on detritus which is already colonized by fungi. The Attine ants (Chapter 8) carry this a stage further and actually culture a specific fungus and feed solely on it. In both cases the fungi are instrumental in transforming the vast amount of energy stored in leaves into forms more acceptable to, and digestible by, the animal.

Traditionally the main function of saprophytic fungi in nature has been considered to be their decomposition of plant and animal remains into simple organic substances which are then recycled by green plants. But as in terrestrial ecosystems, perhaps equally important but certainly less widely appreciated, is the fact that many aquatic animals depend on the accompanying increase in fungal biomass as a rich food source.

The fungal flora of leaves in streams is again very different from those which fall into stagnant ponds. Decaying leaves dredged from the bottom of these are characteristically black in colour. If they are washed and kept damp, but not submerged, in a Petri dish for a few days in the light, clusters of helicoid spores develop on the surface. These belong to Fungi Imperfecti in the genera *Helicoon* and *Helicodendron* (Fig. 6–7). Here, one of the major necessary environmental factors seems to be near anaerobic conditions, rather than well-aerated water. The mycelium grows slowly in the submerged leaves under such conditions but spores are only formed in the gas phase, such as might occur when the pond dries out, especially around the margins, to expose the leaves. The helicoid spores are unwettable and trap air within the coil. If forced below water, they bob up to the surface again like a float. When the pond fills up again they are thus in a good position to colonize fresh leaves as they fall in. Again these fungi are not entirely confined to this habitat. They may also be found under subaerial conditions on very wet, rotten wood.

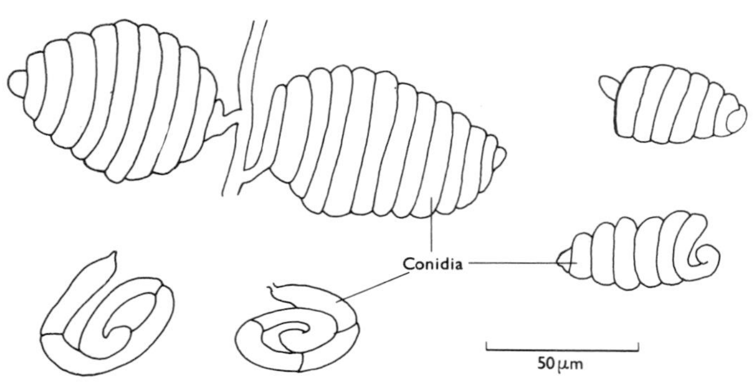

Fig. 6–7 Helicoid conidia.

Each year tremendous numbers of tree leaves are carried to and deposited in the sea by wind, rivers and other agencies. Very little attention has been paid to the fungi which decompose these leaves, although it is well known that fungi attack a wide spectrum of plant and animal detritus in the sea. From the results of one recent study in Canada one is led to conclude that further investigation of the fungal colonizers would be most profitable. Laurel leaves (*Prunus laurocerasus*) were submerged to 1.5 m depth in the sea and collected after 1, 2 and 4 weeks. They were placed in Petri dishes and covered with sterilized filtered sea water. Any fungal hyphae which appeared were isolated to a sea water–glucose agar. One suitable formulation is given in Table 4. The most common fungus to occur on these leaves was a previously undescribed species of Mastigomycete, *Phytophthora vesicula*. It was present on over 70% of the leaves and the common marine Fungus Imperfectus, *Zalerion maritima*, was present on almost 60%.

Attention has already been drawn to the importance of environmental factors in determining which fungi shall colonize any particular substratum. These last three situations, well-aerated streams, stagnant ponds and the sea, further help to stress this point with regard to changes from the normal acrial environment of leaf litter to the aquatic, changes from fresh water to marine, and changes in levels of aeration within freshwaters. The importance of another environmental variable, temperature, will be considered in the next chapter.

7 Thermophilous (thermophilic) Fungi

7.1 Temperature and growth

Temperature, just as moisture, plays a cardinal role in the growth of fungi. The majority of fungi have growth optima between 20 and 30°C and grow between the limits of 5 and 40°C. These are considered as mesophiles. Microbiologists have long recognized the existence of micro-organisms that can grow at temperatures well above those associated with the growth of mesophiles. Thermophile is a convenient term to use for any such heat-loving micro-organism and is generally used for those that grow above 50°C. It should however always be borne in mind that there is no clear dividing line between thermophiles and mesophiles and indeed psychrophiles, which grow below the limits of mesophiles, but rather that there is a continuous spectrum of temperature growth responses.

The best known thermophiles are the blue-green algae of geysers and hot springs, such as those in the Yellowstone National Park, Wyoming, U.S.A., but thermophily is also well-developed in the Actinomycetes, Bacteria and to a less extent in the Fungi. None of the fungi grow above 60°C whereas some of the other organisms grow at much higher temperatures, often 10–15°C above this maximum. The bacterium, *Thermus aquaticus*, isolated from hot springs, will grow at 79°C. The concept of thermophily to be adopted is thus one of degree within a group.

7.2 Thermophilous fungi

Thermophilous fungi are best represented in microbial populations of aerobic or semi-aerobic self-heating organic materials, such as garden and mushroom composts, hay and grasses, manure piles, wood-chip piles, coal-spoil tips and stored moist grain. These are man-made environments but other habitats include nesting material of birds, herbivore dung and soils. There is no evidence as yet, although the possibility exists, that they are more common in the Tropics than elsewhere. There is some evidence that an extensive fungal flora of thermophiles exists in heated soils of volcanic regions.

It is useful to define what we mean by thermophiles as applied to fungi. Many definitions have been attempted but perhaps the best is that given by COONEY and EMERSON (1964). A thermophilous fungus is one that has a maximum temperature for growth at or above 50°C and a minimum for growth at or above 20°C. They base this definition on the evidence that there are a few fungi that grow at temperatures above 50°C and that some of these are unable to grow at ordinary room temperatures of about 20°C. Thus, for example, *Chaetomium thermophile* grows from 27–58°C (Fig. 7–1). This is a purely arbitrary but useful working definition as it excludes thermo-tolerant mesophiles, which can be defined as having minima well below 20°C and maxima near 50°C. For example, *Aspergillus fumigatus* grows from 12–52°C. Such a definition attempts to distinguish between those that merely tolerate, or survive at, the higher temperatures and those that thrive there. Most thermophilous fungi have growth optima between 45 and 50°C and show astonishingly rapid mycelial spread at these temperatures.

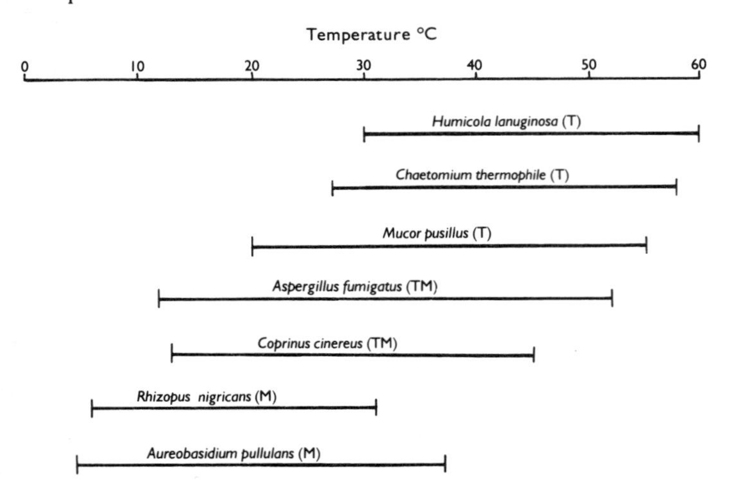

Fig. 7–1 Temperature-growth range of common thermophiles (T), thermo-tolerant mesophiles (TM) and mesophiles (M).

Less than 30 species of thermophilous fungi have so far been described. *Mucor* is the only genus within the Zygomycetes which has a representative amongst these in *M. pusillus*. No thermophilous Loculoascomycetes or Basidiomycetes have been encountered. With the exception of *M. pusillus*, they are all either Euascomycetes or Fungi Imperfecti. The key to their successful isolation is not the use of special culture media, as they will grow on a great variety of standard culture media, but by using incubation temperatures of 45–50°C. If fragments of any appropriate material, such as composting debris, are plated out onto nutrient agar and incubated at 45–50°C, they are sure to develop colonies of thermophilous fungi. Cooney and Emerson prefer to use yeast–starch or yeast–glucose agar (Table 5). These tend to dry out much less rapidly at the higher incubation temperatures than do other media, such as potato dextrose agar. The prevention of desiccation in cultures at such temperatures is always a problem. Small Petri dishes (6 cm diam.) are preferred. The smaller surface area of the agar reduces its rate of desiccation. It is helpful to invert the dishes and to place them in the incubator inside some form of damp chamber, such as a glass dish lined with damp paper towelling.

Table 5 Constituents of yeast–starch agar and yeast–glucose agar.

Yeast–starch agar	
Yeast extract	4.0 g
Potassium phosphate (K_2HPO_4)	1.0 g
Magnesium sulphate ($MgSO_4 . 7H_2O$)	0.5 g
Soluble starch	15.0 g
Agar	15.0 g
Distilled water	1.0 l

Yeast–glucose agar
As above, but with 20.0 g glucose replacing the starch.

7.3 Composts

One of the characteristic features of any composting materials, such as chopped straw or leaves and self-heating hay, is the development of a high temperature phase. The heating is caused by the very active metabolism of the micro-organisms present. The close packing of the materials prevents the rapid dissipation of the heat. Chopped straw can be induced to heat by moistening with a dilute nitrogenous solution, such as ammonium nitrate, and stacking loosely in a large wooden or concrete bin (1 × 1 × 1 m) or in a somewhat larger compound of straw bales. Both the water and the nitrogen source are essential for the induction of the heat phase. The walls of the bin or the straw bales act as insulators. In such composts, the temperature rises quickly in the centre to reach 60–72°C by the fifth or sixth day and is maintained there for two or three days before it begins slowly to fall. It may remain above 40°C for 3–4 weeks before it finally falls to ambient.

In the straw there are many soluble carbohydrates and others, such as starch, which readily become available to microorganisms as soon as it is moistened. This, and the added nitrogen source, stimulates rapid microbial growth. Mesophiles develop initially and raise the temperature to levels where thermophiles are stimulated. Their metabolism raises the temperature to above that which mesophiles can tolerate and these latter are either killed off or persist only around the edges, where the heating is less intense and may not reach 40°C. Thermophilous fungi are active up to 55–60°C. Above this level thermophilous Actinomycetes and Bacteria are left to take over and raise the temperature to its peak. Central areas where the temperature exceeds 60°C are thus devoid of fungi at this time. After the peak the numbers of Actinomycetes and Bacteria decline and thermophilous fungi recolonize from cooler marginal areas and maintain the temperature above 40°C for quite a prolonged period. This is the period of maximum activity of the markedly cellulolytic *Chaetomium thermophile* and *Humicola insolens* and some associated secondary sugar fungi, such as *Thermomyces lanuginosus* (Fig. 7–2). These are much less active, although they may be present with other fungi, including *Mucor pusillus* and *Aspergillus fumigatus*, before the peak heating. As the

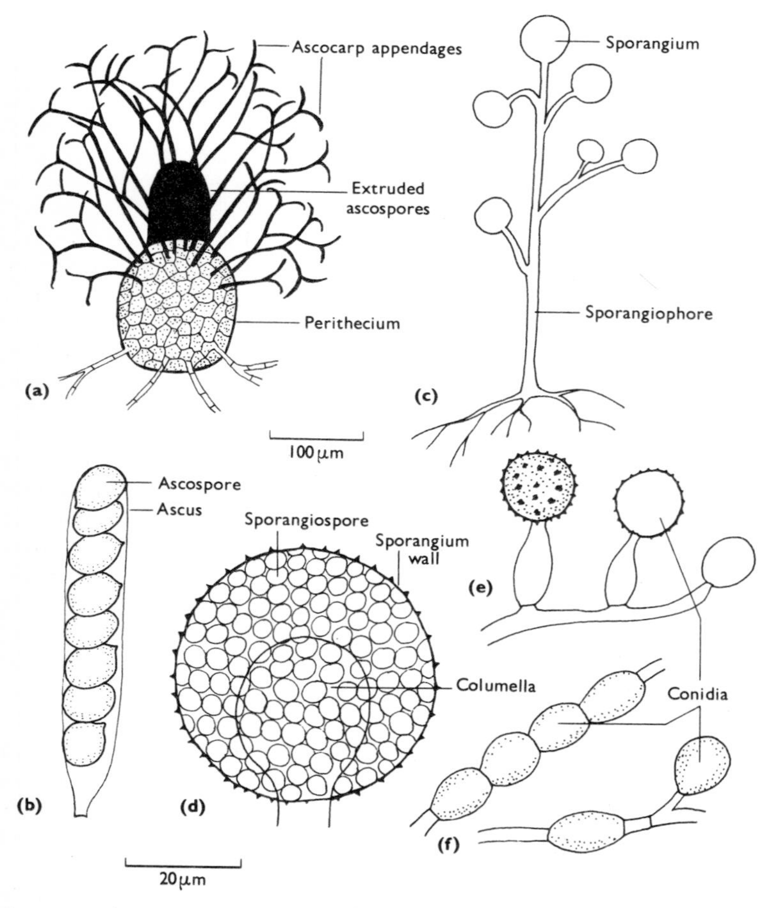

Fig. 7–2 Thermophilous fungi. (**a**) perithecium and (**b**) ascus of *Chaetomium thermophile*, (**c**) sporangiophore and (**d**) sporangium of *Mucor pusillus*, (**e**) conidia and conidiophore of *Thermomyces lanuginosus* and (**f**) conidia of *Humicola insolens*.

temperature drops to below 35°C, several mesophiles recolonize. The most notable of these are a number of small *Coprinus* sp., such as *C. cinereus* (Fig. 4–3). Some of these are capable of utilizing both cellulose and lignin.

In broadest outline the succession is similar to the coprophilous one with an initial, but shorter, Zygomycete phase, followed by Ascomycetes and Fungi Imperfecti and terminating with a Basidiomycete phase. Although temperature is undoubtedly the major environmental factor determining the sequence, the resemblance to the coprophilous fungal succession would suggest that additional environmental factors are also operative. Herbivore dung has an initially higher nitrogen level than the ingested material. It was suggested that this influenced the succession. Similarly the addition of ammonium nitrate to the straw might also help to determine the sequence in composts. The moisture content of the composting straw, like dung, is also higher and subject to less marked fluctuations than is naturally decomposing straw. This, again, would favour more rapid decomposition. It is the complex interaction of such ecological factors that determine which fungi shall succeed on a particular substratum.

In normal garden composts and in many municipal composting systems, three factors may limit the growth of thermophilous fungi. These are excessively high temperatures, acidity and low oxygen levels. For example, adding abundant grass cuttings to garden compost heaps may cause a rapid rise in temperature to well over 70°C and cause compaction which severely reduces the inward diffusion of oxygen. Although fungi are obligate aerobes, several species of thermophilous ones have been shown to produce significant growth in oxygen tensions as low as 0.2–1.0%. In this particular example temperature would be the major limiting factor. The response of thermophilous fungi to pH varies with the species. Some will grow uniformly over a pH range of 4–8 but others are much more sensitive and most isolated from composts tend to have a neutral or alkaline pH optimum and in this respect are like many coprophilous and pyrophilous fungi.

7.4 Mushroom compost

There are many species of edible fungi in this country notably in such genera as *Agaricus, Boletus, Lepiota, Lepista* and *Morchella*. None of our wild species have been successfully cultivated commercially. The cultivated mushroom has sufficient morphological and physiological differences from other species of *Agaricus* to justify it being considered as a separate species, *A. bisporus*. Mushroom cultivation was started in the early part of the seventeenth century by French horticulturalists and, although initially an outdoor crop, by 1850 mushroom growing was a thriving industry in the quarried caves of the Paris region. At that time, and until quite recently, mushroom growing was more or less a gamble but now, with experience and a more thorough knowledge of its biology, excellent crops are produced with certainty on an enormous scale throughout Europe and N. America.

It is usually grown on a horse manure and straw mixture which must undergo a certain amount of decomposition before the mushroom will grow well in it. The mixture is first allowed to 'ferment' in compost heaps between 1.2–1.8 m high and any convenient width and length. Care is taken to see that the compost is uniformly moistened. After a few days, the compost becomes hot as a result of microbial thermogenesis, as in straw composts. The temperature may exceed 75°C. *A. bisporus* is a mesophile so that fresh horse manure is unsuitable because of the heat generated as fermentation occurs. To produce a good compost it is essential to maintain good aeration, adequate retention of heat and a uniform moisture content. The height of the heap is important. The higher the heap, the greater the retention of heat, but the poorer the aeration of the lower layers. As the temperature rises, the edges may dry out as water vapour is given off. The centre and bottom may become anaerobic. The heaps are thus turned several times. Various 'conditioners', such as gypsum, may be added. Gypsum improves the final texture of the compost. It prevents over-compacting and produces a short, fibrous, well aerated, brown-coloured compost of pH 6.5–7.5.

When the mixture is sufficiently composted, it is made up into flat or ridged beds or filled into trays inside the mushroom house or cave. Flat beds are about 90 cm wide and in these the compost is spread evenly and compacted to a depth of 15–18 cm. The compost then heats again to between 54 and 63°C. The temperature may have to be kept down by ventilation. This so-called 'sweating-out' period lasts for 7–10 days during which time the temperature gradually falls. Several thermophilous fungi, especially *Chaetomium thermophile* and *Humicola insolens*, are particularly active in the compost at this time in addition to Actinomycetes and Bacteria. The effect of this temperature rise is by no means fully understood but, judging by the substantial increase in yield over composts which have not undergone this heat stage, it is important in preparing the compost for *A. bisporus*. The most plausible explanation is that controlled microbial decomposition during this phase produces nutrients, especially proteins, which are necessary if the mycelium of *A. bisporus* is to grow. At the same time any residues of ammonia or amines left at the end of the first phase of thermogenesis are decomposed or converted. They are toxic to the mycelium of *A. bisporus* but not to 'weed' moulds which would subsequently colonize the compost and compete with *A. bisporus* for nutrients. This phase also effectively pasteurizes the inside of the compost and at some stage the air temperature around the compost is raised to 60–62°C for a short period to complete the process of pasteurization, killing any spores on the surface as well as pests such as mites, other insects and nematodes.

When the temperature finally falls to below 30°C the compost may be 'spawned'. Most growers prefer to wait until the temperature falls to 21–23°C. Spawn is a pure culture of *A. bisporus* mycelium and spawning consists of inoculating beds evenly, at about a 25 cm spacing, by inserting walnut-size pieces of spawn into the compost at a depth of 1.5–2.5 cm or by thoroughly mixing the compost with a granular spawn produced by growing the fungus on sterilized rye or wheat grains with added chalk to maintain alkalinity. By this time, the other microorganisms present have removed all the readily available carbohydrates so that conditions are conducive for the secretion of cellulases by the strongly cellulolytic *A. bisporus*. Being a massive mycelial inoculum, it takes over the bed completely to the exclusion of all other fungi. Hyphae permeate the compost and it is said to have 'run'. After 2–3 weeks the beds are 'cased'. The purpose of this casing is to stimulate the mycelium to 'crop' i.e. produce basidiocarps. It consists of covering the beds with a thin layer (2.5 cm) of unsterilized soil or soil-like substances, such as peat-chalk or peat-vermiculite mixtures. The transition from the vegetative to the reproductive phase takes place in the compacted casing layer. The precise trigger for this switch is not known but numerous suggestions have been made. In contrast to the compost itself, the casing layer is very low in nutrients and it may be the lack of particular nutrients which bring about basidiocarp production. The growing mycelium in both the casing layer and the compost produces volatile metabolites such as ethanol, acetaldehyde, ethyl acetate and carbon dioxide. These accumulate in the casing layer and are selective for a highly adapted microflora, especially the bacterium *Pseudomonas putida*. One suggestion is that it triggers off basidiocarp production by releasing ferrous iron, which is essential for the development and growth of basidiocarps, by releasing it from organic complexes in the peat. Another suggestion is that the bacteria may be responsible for removing inhibitors, possibly volatiles, of basidiocarp formation produced by the mycelium itself. What is clear is that in the absence of a casing layer, or if the casing is sterilized first, little or no cropping occurs. After a further 2–4 weeks, basidiocarps begin to appear. Several flushes may develop over the following 4–12 weeks to give a total yield of 7–10 kg m^{-2} of bed before the compost is exhausted. The yield also depends on the maintenance of certain environmental conditions. For about 4 weeks after spawning, the air temperature is maintained between 15 and 21°C and then gradually reduced to between 13 and 15°C. The reason for this is that the higher temperature favours rapid mycelial growth and the lower is more suitable for basidiocarp formation. The latter also requires a high humidity so that the relative humidity is maintained between 88 and 90%. Ventilation is also important. The accumulation of carbon dioxide to near 5% by volume around the beds or trays can inhibit mycelial growth and lower concentrations, around 1%, will produce deformed basidiocarps with long stalks and small caps.

There are very many modifications of this complicated process. These are discussed in detail by CHANG and HAYES (1978), SINGER (1961) and

RAMSBOTTOM (1953) gives a short history of mushroom cultivation. The major difference from ordinary composting, in terms of the occurrence of fungi, lies in the manipulation of the final Basidiomycete phase when the bulk of the cellulose and lignin is utilized. The idea is to add such a large mycelial inoculum that it has sufficient competitive ability to take over the compost completely and replace the small Coprini. The latter, however, do often occur in mushroom beds in small quantities and are either regarded as harmless 'weed' fungi or as a good sign in that if the compost is favourable to their growth, it will also be so for *A. bisporus*.

7.5 The cultivation of other Basidiomycetes

Although *A. bisporus* is the only fungus cultivated in Europe and N. America, other species are grown elsewhere. *Volvariella volvacea*, the Padi straw mushroom, is extensively grown in India and the Far East. Bundles of well-soaked rice straw are laid in beds outside and inoculated with either a natural spawn from another bed or a pure culture spawn. More primitively, infection is left to chance. Given warm weather (above 21°C) and high humidities, basidiocarps are produced in 2–3 weeks. A similar species, *V. speciosa*, is common on sodden baled straw and cut grass by the roadside in this country.

Lentinus edodes, Shiitake, has been known and highly rated as a food crop in China and Japan for over 2000 years. Unlike *Agaricus* and *Volvariella*, it is a wood-inhabiting Basidiomycete and grows on dead wood of *Shiia*, *Quercus*, *Castanea* and other hardwoods. In its cultivation, logs, 5–15 cm diam., are cut from coppiced wood into 1 m lengths. They are soaked and the bark pounded. A spore suspension or mycelial inoculum is introduced into holes in the logs which are then stacked, each at an acute angle to the ground, in a clearing at the margin of the forest. After 5–8 months, the mycelium of *Lentinus* will have completely permeated the logs. They are then moved to a more shaded and moist environment and stood in an almost upright position, in rows against bamboo supports. This occurs in winter and in the following spring, they begin to crop. Two crops a year, one in spring and one in autumn, may be produced for 3–5 years. Shiitake is one of the most important forest by-products of Japan. The well established cultivation of Shiitake suggest the possibility of attempting to grow other common edible wood-inhabiting fungi, such as *Agrocybe aegerita* and *Pleurotus ostreatus*. There is ample scope for experimentation here.

7.6 Thermophilous fungi in birds' nests and grain storage silos

Birds' nests appear to be another favourable habitat for the development of thermophilous fungi. The body temperature of birds is relatively high, usually between 40–43°C. Their nests contain suitable organic materials and the birds warm these while using the nests, especially for incubation of their eggs. The fungi are present on the nesting materials, as spores, before it is used for the construction of the nest and only grow under favourable moisture and temperature regimes. The number of thermophiles present and their frequency varies considerably with the species of bird. Sandmartins' nests are exceptionally low whereas the nests of blackbirds, thrushes and hedgesparrows are rich in thermophiles. The sandstone holes in which sandmartins nest may be too cool for their development.

Another environment where organic materials are kept in sufficiently large masses to be self-insulating are commercial grain storage silos. If aeration is adequate in these and the moisture content rises above 14% in any part, thermophilous fungi may develop in addition to storage fungi. Hot spots develop initially and later lead to a more general self-heating which spreads rapidly.

7·7 Fungi in cold environments

Fungi can also be found growing in very cold environments, such as the Arctic and Antarctic, where the temperature is below freezing point for very long periods and the air temperature rarely exceeds 5–10°C. It is not possible to give an absolute temperature below which fungi will not grow. Growth at very low temperatures may be imperceptibly slow and the duration of exposure is most important. There are, however, only a few records of fungal growth below −6°C. A yeast has been recorded as growing at −34°C.

Tundra vegetation provides numerous substrata for fungi within the Arctic. Any fungi found there must be able to tolerate extreme cold (−50°C or even below) but we do not know the temperatures at which such fungi make significant growth. They may only grow in those relatively short periods when the temperature rises to between 0 and 15°C and for the rest of the year remain inactive. The number of species present in the Tundra is smaller than in Temperate regions, but the fungal flora is nevertheless very diverse. All classes of fungi are represented. For instance, the whole range of coprophilous fungi are present. Too little is known of their ecology to state whether they are true psychrophiles actually growing better at these low temperatures and with growth optima below 10°C or whether they are merely cold-tolerant mesophiles.

Many of the fungi of another, but man-made, cold environment, refrigerated foods, are cold-tolerant mesophiles. The majority of these grow best at normal temperatures, but are able to grow, albeit relatively slowly, at lower temperatures than most fungi. The common mesophile *Cladosporium herbarum* will grow on meat at −6–0°C and grows quite profusely at 2°C. The habit of low temperature storage of foodstuffs may select strains of fungi that can grow at temperatures below freezing.

8 Saprophytes Put to Use

8.1 Fermented drinks and foods

Since time immemorial man has used saprophytic fungi. The knowledge that yeasts convert sucrose to alcohol and carbon dioxide is fundamental to brewing and bread baking. In brewing, alcohol is the important product but carbon dioxide is a valuable by-product and is marketed as 'dry-ice'. Most yeasts do not secrete amylases and in the fermentation of starchy materials, filamentous fungi, such as *Mucor*, *Rhizopus* and *Aspergillus*, are used to hydrolyze these to sugars. The sugars are then fermented to alcohol by yeasts. Kaffir beer, drunk by natives of Malawi, is produced by fermenting *Sorghum* grains. Species of *Aspergillus* and *Mucor* convert the starch into sugar. Fermentation is then carried out by yeasts. In addition to alcohol, the beer contains appreciable quantities of riboflavin and nicotinic acid. Natives who regularly drink the beer do not suffer from pellagra, which is usually associated with the lack of these vitamins in their basic maize meal diet. In other fermented drinks, such as those produced by the 'Tea-fungus' and the 'Ginger-beer plant', two micro-organisms are also involved, a bacterium and a yeast. Both these fermentations produce a mildly acidic, alcoholic drink. The former also has certain antibiotic properties.

Fermented foods are less well-known here than they are in many eastern countries. One such food, Tempeh, is regularly eaten by millions of people in Indonesia, New Guinea and Surinam. It is a product of soybeans. The beans are soaked overnight, the seed coats removed and the dehulled beans boiled. After draining and cooling, they are inoculated with an appropriate strain of *Rhizopus*. The beans are then tightly packed and after incubation at 31–32°C for 24 h, they become completely bound together by the white mycelium of *Rhizopus*. The whole is then cut into slices, dipped in salt water and fried in vegetable oil for breakfast. In this fermentation it is not the production of amylolytic enzymes which is important but the production of proteolytic and lipolytic ones. These enhance digestion of the soybeans and make them more palatable by invariably adding flavour. Over half of the proteins present are hydrolyzed to amino acids and fats are converted to fatty acids. There is again an increase in vitamin content, especially riboflavin and nicotinic acid.

The brewing of Shoyu or soy sauce has taken place in Japan for over 1000 years and was probably introduced with Buddhism from China around A.D. 550 and the change to a vegetable diet. In this process, washed, autoclaved soybeans are mixed with an almost equal volume of roasted crushed wheat. The latter gives flavour and aroma. The mixture is inoculated with *Aspergillus oryzae*, growing on cooked rice, and incubated at 25–35°C for 3–4 days. As fermentation proceeds it has to be cooled to prevent the temperature rising above 40°C. *A. oryzae* is a mesophile. It is them placed in tanks or deep vessels with an equal quantity of brine for 3–4 months. During this period further fermentation by lactic acid bacteria and a yeast occurs. The final mash is pressed into a cake and the extract obtained is Shoyu. It is widely used for flavouring and seasoning. Its flavour is mainly due to glutamic acid. HESSELTINE (1965) gives details of these and a large number of other fermentation products made in the Far East. We are more familiar with the flavours fungi impart to ripening cheeses, such as Camembert, Brie or blue-veined Stilton.

8.2 Acids, antibiotics, steroid conversions and vitamins

In many fungal fermentations, acids may be the final product not alcohol. Citric acid, also used as a food flavouring, is a fermentation product of many fungi and is produced commercially from *Aspergillus niger*. The main essentials for its production, other than a good strain of *A. niger*, are a high initial sucrose concentration, about 12–15%, a low nitrogen source, usually ammonium nitrate, an acid culture medium, pH 1.5–3.5, and a low air flow to retain carbon dioxide. The *A. niger* is allowed to form a mycelial mat on the surface of shallow vats of nutrient solution. The acidification, with a mineral acid, is to prevent growth of any bacterial contaminants which might arise under these conditions. After 8–10 days about 70% of the sugar is converted to citric acid. Several other acids, including gluconic and fumaric, are produced in this way.

Fungi are also notorious for secreting complex metabolites into their environment. These include, in addition to enzymes, vitamins and antibiotics. These latter, one assumes, are substances produced by fungi which will inhibit or combat certain competitor micro-organisms in their natural environment. The best known, most documented and most useful to man of these are the penicillins. There are many other useful ones, such as ceporin and griseofulvin. These are produced by fermentation processes. Griseofulvin is produced by a number of species of *Penicillium*, especially *P. patulum*. It is most unusual in that its effects are exerted specifically against chitin-walled fungi. It is thus not effective against either the Oomycetes and yeasts or Actinomycetes and Bacteria. Fungal dermatophytes, growing on skin, nails and hair, are particularly sensitive to it. It has strong affinities for keratinous substrata and has been developed as an oral antibiotic against dermatophytes. It passes into the bloodstream and accumulates in keratinous zones of the body, where it has a fungistatic rather than a fungicidal effect on any fungal pathogens.

The ability of fungi to convert one substrate into another is used in the production of medically important steroids. The physiological activity of a steroid molecule often depends on the possession of a particular substituent at a specific part of the molecule, such as a hydroxyl group on carbon atom-II. The addition of such a group may be difficult by purely chemical means. However some fungi can do this extremely efficiently. Many types of steroid transformation can be carried out by different fungi. These include, in addition to hydroxylation, ketone formation, cleavage of carbon side chains and various other reactions.

Increased levels of vitamins were noted in Tempeh. One of the richest sources of the vitamin B complex is dried yeast. Riboflavin, one of this complex, is produced commercially by yeast fermentation. 'Food Yeast', a rich source of protein and vitamin B, is manufactured from large scale cultures of the yeast, *Candida utilis*. The yeast is grown on bulk cheap carbohydrates, such as molasses or potato starch, with added ammonium nitrogen. Crude hydrocarbons from the petroleum industry are another bulk cheap organic carbon source. There is every prospect that useful compounds can be produced by fungal fermentation of these or that *Candida* and other yeasts can be grown on them to produce single cell protein as a non-agricultural means of producing food.

8.3 Fungi as food

The commercial cultivation of mushrooms has already been considered (section 7.4). Many of the larger fungi of our woodlands and pastures are edible and are considered as delicacies by gourmets. The minority are poisonous, and a few deadly so. Toadstools have been eaten from the earliest of times. The British appear more prejudiced against them than do the Continentals. In several Continental countries many fungi are sold in the markets and some towns have special fungus markets. RAMSBOTTOM (1953) gives an excellent account of edible and poisonous fungi. He points out that there are in this country a number of species which have always had a good reputation as being edible. These are easy to identify and cannot readily be confused with poisonous ones. For those interested in mycophagy it is advisable to recognize some of these and not to experiment with others.

Man is not the only animal to use fungi as a food source. The Attine ants (*Atta colombria*) culture one of the Basidiomycete Agaricales in 'fungus gardens' in their nests. They forage for and cut up leaves of living plants, place these in their 'fungus gardens' and inoculate them with mycelial fragments, much as a mushroom grower would spawn his compost. They also deposit faecal droplets on the cut fragments. The droplets contain significant quantities of ammonia and amino acids which are beneficial in producing nitrogen to initiate the growth of the fungus. Their benefit is short lasting as they are present in very small volume in relation to the quantitites of leaves used and are in any case rapidly absorbed and utilized. The bulk source of nitrogen in the leaves is protein and, if the fungus is to use the leaves as its sole substratum, these must be hydrolysed to provide additional sources of nitrogen. The faecal droplets also contain proteases which catalyse the hydrolysis of these leaf proteins. The proteases are of fungal origin and one important role of the ants appears to be to acquire and accumulate these as they feed on the fungus. They move them from a region in the nest where the fungus is in a state of rapid growth and the enzymes are in ample supply to a site of inoculation, the newly added leaves, where they are in short supply. This favours the rapid and continuous growth of the fungus. The ants avoid digesting the fungal enzymes through the simple expediency of not secreting any digestive proteases of their own. The ants obtain their own nitrogen supplies in the form of amino acids from the ingested fungal cytoplasm. The fungus is cellulolytic and it indirectly contributes its cellulose degrading ability to the ants. It converts the cellulose in the leaves into such fungal carbohydrates as trehalose and glycogen and the polyhydric alcohol mannitol which the ants can utilize. Thus the ants can tap the virtually inexhaustible cellulose supply in their environment. Once having established the fungus the ants use it as their sole source of food. The Attine ant–fungus association is an obligate symbiosis for both. The success of these ants in the New World Tropics must be attributed to their leaf-cutting, fungus-growing activities.

8.4 Control of fungal pathogens

One of the few methods of biological control of plant pathogens that has proved successful in practice is the control of *Fomes annosus*, which causes a serious root- and butt-rot of pines and other conifers. It is a Basidiomycete and one means of spread is by air-borne spores, which colonize still living stumps of recently felled trees. It grows down into the roots and can spread by root contact to healthy trees. Such infected stumps form a large inoculum for infection of surrounding trees and any replants.

Many attempts at control have been devised. These include painting stump surfaces with chemical protectants, such as creosote and ammonium sulphamate, to reduce infection. *F. annosus* has many competitors in its colonization of pine stumps. The most successful of these is *Peniophora gigantea*, another Basidiomycete. As a saprophyte, it is a very active decomposer of cellulose and lignin. If spores of both are applied to cut stumps, *P. gigantea* is the better competitor and checks infection by *F. annosus*. This has been developed as a control method (RISHBETH, 1963). A suitable spore inoculum of *P. gigantea* is now produced in the form of a concentrated suspension sealed in sachets. The spores from these are dispersed in water and applied to stumps immediately after felling. The fungus quickly becomes established leading to rapid breakdown of stump and root tissues and to the exclusion of *F. annosus*. Some form of stump treatment is mandatory both at thinning and clear-felling in Forestry Commission conifer plantations.

Further Reading

CHANG, S. T. and HAYES, W. A. (1978). *The Biology and Cultivation of Edible Mushrooms*. Academic Press, New York and London.

CHRISTENSEN, C. M. (1965) *The Molds and Man*. University of Minnesota Press, Minneapolis.

COONEY, D. G. and EMERSON, R. (1964) *Thermophilic Fungi*. W. H. Freeman & Co., San Francisco and London.

COMMONWEALTH MYCOLOGICAL INSTITUTE (1968). *Plant Pathologist's Pocketbook*. Commonwealth Mycological Institute, Kew, Surrey.

DEVERALL, B. J. (1969) *Fungal Parasitism*. Studies in Biology, no. 17. Edward Arnold, London.

GREGORY, P. H. (1973). *The Microbiology of the Atmosphere*. Leonard Hill (Books) Ltd., London and Interscience Publishers, Inc., New York.

HARPER, J. E. and WEBSTER, J. (1964). *Trans. Br. mycol. Soc.*, **47**, 511–30.

HESSELTINE, C. W. (1965). *Mycologia*, 57, 149–97.

HORA, F. B. (1959). *Trans. Br. mycol. Soc.*, **42**, 1–14.

INGOLD, C. T. (1960). *Dispersal in Fungi*. Clarendon Press, Oxford.

INGOLD, C. T. (1971). *Fungal Spores*. Clarendon Press, Oxford.

INGOLD, C. T. (1975). An illustrated guide to aquatic and water-borne Hyphomycetes. *Freshwater Biological Association Scientific Publication*, no. **30**.

INGOLD, C. T. (1979). *The Nature of Toadstools*. Studies in Biology, no. 113. Edward Arnold, London.

JACKSON, R. M. and RAW, F. (1966). *Life in the Soil*. Studies in Biology, no. 2. Edward Arnold, London.

LANGE, M. and HORA, F. B. (1965). *Collins Guide to Mushrooms and Toadstools*. Collins, London.

RAMSBOTTOM, J. (1953). *Mushrooms and Toadstools*. The New Naturalist, no. 7. Collins, London.

RICHARDSON, M. and WATLING, R. (1968). *Bull. Br. mycol. Soc.*, **2**(**1**), 18–43.

RICHARDSON, M. and WATLING, R. (1969). *Bull. Br. mycol. Soc.*, 3(2), 86–8, 121–4.

RISHBETH, J. (1963). *Ann. appl. Biol.*, **52**, 63–77.

SCOTT, G. D. (1969). *Plant Symbiosis*. Studies in Biology, no. 16. Edward Arnold, London.

SINGER, R. (1961). *Mushrooms and Truffles*. Leonard Hill (Books) Ltd., London and Interscience Publishers, Inc., New York.

WEBSTER, J. (1970) *Introduction to Fungi*. Cambridge University Press, London.

WILKINS, W. H., ELLIS, E. M. and HARLEY, J. L. (1937). *Ann. appl. Biol.*, **24**, 703–32.

WILKINS, W. H. and HARRIS, G. C. M. (1946). *Ann. appl. Biol.*, 33, 179–88.